Organic
Vegetable Production
Manual

Technical Editor
Milton E. McGiffen, Jr.

Publication Coordinators
Jeri Ohmart and David Chaney

2011

University *of* **California** Agriculture and Natural Resources

Funding for writing this publication provided in part by the California
Department of Food and Agriculture *Buy California Initiative*.

cdfa

CALIFORNIA DEPARTMENT OF
FOOD AND AGRICULTURE

To order or obtain ANR publications and other products, visit the ANR Communication Services online catalog at http://anrcatalog.ucdavis.edu or phone 1-800-994-8849. You can also place orders by mail or FAX, or request a printed catalog of our products from

University of California
Agriculture and Natural Resources
Communication Services
1301 S. 46th Street
Richmond, California 94804-4600

Telephone 1-800-994-8849 or 510-665-2195
FAX 510–665–3427, E-mail: anrcatalog@ucdavis.edu

Publication 3509
ISBN-13: 978-1-60107-557-4
Library of Congress Control Number: 2011935245

Cover photograph credits: Richard F. Smith (front cover); iStock photo (back cover top); Kathy Keatley Garvey (back cover bottom).

To simplify information, trade names of products have been used. No endorsement of named or illustrated products is intended, nor is criticism implied of similar products that are not mentioned or illustrated.

UC PEER REVIEWED This publication has been anonymously peer reviewed for technical accuracy by University of California scientists and other qualified professionals. This review process was managed by ANR Associate Editor for Vegetable and Agronomic Cropping Systems Jeff Mitchell.

⊕ Printed in the United States of America on recycled paper.

4m-pr-10/11-WJC/RW

◾ Contents

■ Preface and Acknowledgments

Organic Vegetable Production Manual provides detailed information for growers on how to farm vegetables organically. It addresses production issues, economics, weed and disease management, the conversion process, and organic certification and registration, all essential topics for succeeding in a highly competitive marketplace.

The authors of this volume had initially planned to include a chapter on insect pests, but scheduling delays on that chapter's manuscript made that impossible. Instead, the authors refer you to other publications in the UC ANR catalog (anrcatalog. ucdavis.edu) and to the wealth of information available online from the UC Statewide Integrated Pest Management Program (www.ipm.ucdavis.edu).

The number of organic farmers and the acreage devoted to organic production have increased. From 1992 to 2003, California's organic vegetable acreage quadrupled to nearly 60,000 acres, generating 2 percent of the total cash income from vegetable production for the entire state. As interest in organic production has increased, so has the need for information. In 2000, the University of California's Division of Agriculture and Natural Resources (UC ANR) produced a series of short publications on organic vegetable production for California growers:

Organic Certification, Farm Production Planning, and Marketing (ANR Publication 7247)

Soil Management and Soil Quality for Organic Crops (ANR Publication 7248)

Soil Fertility Management for Organic Crops (ANR Publication 7249)

Weed Management for Organic Crops (ANR Publication 7250)

Insect Management for Organic Crops (ANR Publication 7251)

Plant Disease Management for Organic Crops (ANR Publication 7252)

Postharvest Handling for Organic Crops (ANR Publication 7254)

This series was an important resource for both organic and conventional growers during the early stages of implementation of the new national organic program.

Other information resources related to (if not directly applicable to) organic vegetable production are available from UC ANR. Crop-specific integrated pest management (IPM) manuals include *IPM for Cole Crops and Lettuce, IPM for Potatoes*, and *IPM for Tomatoes*. In addition, the UC Statewide IPM Program offers *Pest Management Guidelines* for particular commodities on its Web site (www.ipm.ucdavis.edu/ PMG/crops-agriculture.html). *Pest Notes*, produced by UC IPM mainly for landscapers and home gardeners, may also be of interest (www.ipm.ucdavis.edu/PDF/ PESTNOTES). Organic pest management strategies are addressed in all of these publications.

This new publication, *Organic Vegetable Production Manual*, includes information from that series with substantial revisions and additions to address the changing needs of organic growers, the implementation in 2002 of national organic standards, and the availability of new information.

Organic Vegetable Production Manual incorporates updates to the previous publication series and provides detailed information on production issues, economics, pest control, the conversion-to-organic process, and organic certification and registration in California. It was written with support from the California Department of Food and Agriculture's Buy California Initiative. Note that the content of this publication does not necessarily reflect the views or policies of CDFA or USDA, nor does any mention of trade names, commercial products, and organizations imply endorsement of them by CDFA or USDA. Funding for this project supported a series of workshops on organic production, including a January 18, 2005, session on vegetable production held in Salinas. Much of the material in this publication is based on the presentations given at that workshop. The authors would like to thank Jeri Ohmart of the UC Sustainable Agriculture Research and Education Program and David Chaney for their assistance in organizing the workshop and for their coordination of the many tasks involved in seeing this manual through to publication.

Additional information on organic vegetable production is available from several online sources, including the following:

UNIVERSITY OF CALIFORNIA

Agricultural Issues Center
http://aic.ucdavis.edu/

Davis Agricultural and Resource Economics Cost Studies
http://coststudies.ucdavis.edu

Small Farm Program
www.sfp.ucdavis.edu

Statewide Integrated Pest Management Program
www.ipm.ucdavis.edu

Sustainable Agriculture Research and Education Program
www.sarep.ucdavis.edu/organic

Vegetable Research and Information Center
http://vric.ucdavis.edu/

GOVERNMENT

California Department of Food and Agriculture, State Organic Program
www.cdfa.ca.gov/is/i_&_c/organic.html

USDA Economic Research Service Organic Agriculture Briefing Room
www.ers.usda.gov/briefing/Organic/

USDA National Organic Program
www.ams.usda.gov/AMSv1.0/nop

USDA Sustainable Agriculture Research and Education
www.sare.org

OTHER

ATTRA National Sustainable Agriculture Information Service
https://attra.ncat.org/organic.html

eOrganic
http://eorganic.info

The New Farm Organic Price Report
www.rodaleinstitute.org/Organic-Price-Report

Organic Farming Research Foundation
www.ofrf.org

■ Authors

TECHNICAL EDITOR

Milton E. McGiffen, Jr., Botany and Plant Sciences,
UC Riverside

CONTRIBUTING AUTHORS

David Chaney, DEC Education Services, Corvallis,
Oregon (formerly with the UC Sustainable
Agriculture Research and Education Program)

Richard L. De Moura, Agricultural and Resource
Economics, UC Davis

Calvin Fouche, UC Cooperative Extension,
San Joaquin County

Mark Gaskell, UC Cooperative Extension, Santa Barbara
and San Luis Obispo Counties

Ray Green, formerly Organic Program Manager,
California Department of Food and Agriculture

Tim Hartz, Plant Sciences, UC Davis

William Horwath, Land, Air, and Water Resources,
UC Davis

Louise Jackson, Plant Sciences, UC Davis

Karen Klonsky, Agricultural and Resource
Economics, UC Davis

Steven T. Koike, UC Cooperative Extension,
Monterey and Santa Cruz Counties

W. Thomas Lanini, Weed Science Program, UC Davis

Jeff Mitchell, UC Kearney Agricultural Center, Parlier

Richard F. Smith, UC Cooperative Extension,
Monterey and Santa Cruz Counties

Alexandra Stone, Department of Horticulture,
Oregon State University

Trevor Suslow, Plant Sciences, UC Davis

L. Ann Thrupp, Manager of Sustainability and Organic
Development, Fetzer Winery, Mendocino

Laura Tourte, UC Cooperative Extension,
Santa Cruz County

PUBLICATION COORDINATORS

Jeri Ohmart, UC Sustainable Agriculture Research
and Education Program

David Chaney, DEC Education Services, Corvallis,
Oregon (formerly with the UC Sustainable
Agriculture Research and Education Program)

Organic
Vegetable Production
Manual

Organic Certification and Registration in California

DAVID CHANEY, RAY GREEN, AND L. ANN THRUPP

BACKGROUND

The organic industry has grown significantly during the past 20 years. For the United States as a whole, the total market value (retail sales) of organic food products, including processed products, grew from about $1 billion in 1990 to an estimated $21.1 billion in 2008 (Dimitri and Greene 2002, Economic Research Service-USDA 2011), while the Organic Trade Association estimated 2009 total organic product sales in the United States at $26.6 billion (OTA 2010). California producers have led this trend showing an increase in both numbers of organic farmers and total acreage. From 1992 to 2008, the number of registered organic farms in California more than doubled, from 1,273 to 2,961 (Klonsky 2010). Over the same period organic acreage grew more than tenfold, increasing from 42,000 acres in 1992 to 470,700 acres in 2008 (Klonsky 2010; Klonsky and Tourte 1998, 2002; Klonsky and Richter 2005).

Prior to 1990, few common labeling standards were agreed upon and for the most part there was no regulatory requirement for organic certification. As the market began expanding, organic producers and marketers recognized the need for more uniformity and integrity in their products and they turned to Congress for assistance in developing national standards to govern organics. The result of these efforts was the Organic Foods Production Act (OFPA), passed in 1990 as part of the Food, Agriculture, Conservation, and Trade Act. OFPA mandated that the U.S. Department of Agriculture (USDA) establish an organic certification program for producers and handlers of agricultural products who sell and label their products as being organic. Responsibility for developing and administering a National Organic Program (NOP) was assigned to USDA's Agricultural Marketing Service (AMS). The intent of OFPA was to establish national practice standards governing the production and marketing of certain organically produced agricultural products, to assure consumers that organically labeled foods represented these consistent standards, and to facilitate interstate commerce in fresh and processed food that is organically produced and labeled as such.

California did have an organic labeling law in place starting in the 1970s, but the law had no enforcement component. Concurrent with the development of OFPA, California passed its own law regulating organic farming, the California Organic Foods Act (COFA), which was signed into law in 1990. COFA set a voluntary certification standard with full registration and enforcement components. The California law (COFA) became effective immediately upon signing, but it took more than 10 years for the federal OFPA to be completed and implemented. When COFA was eventually superseded by the federal regulations, the law was rewritten to become the California Organic Products Act of 2003 (COPA).

The process of completing the national rule was complicated and controversial, involving numerous drafts and public input from consumers and the organic industry. USDA issued a proposed rule in 1997, but the final program was not completed until December 2000. The final rule states that all organic producers who gross more than $5,000 per year in retail sales must be certified through an accredited certification agency. These certifying organizations, both public and private, act as third-party agents of the NOP to legally verify that production and handling practices meet the national standards. Enforcement of the NOP began in October 2002, and AMS immediately began to accredit certification agencies to assist with its implementation.

In addition to accreditation of certifiers, the NOP is also responsible for ensuring that the purposes of the OFPA are accomplished, determining the equivalency of foreign programs for imports into the United States, participating in the development of international standards, coordinating enforcement activities with other agencies, conducting the petition process for materials review, and providing administrative support for the National Organic Standards Board (NOSB). The NOSB is a 15-member committee established as part of OFPA to help the secretary of agriculture develop standards for substances to be used in organic production and to advise the secretary on other aspects of implementing the program. The current board includes four farmers/growers, two handlers/processors, one retailer, one scientist, three consumer/public interest advocates, three environmentalists, and one certifying agent.

STATES AND THE NOP

State governments can serve two functions in relation to the NOP: they can either choose to implement a State Organic Program (SOP) or become an accredited certifying agent, or they can choose to do both. An SOP is primarily an enforcement program and is not involved in certifying growers or handlers. Once approved as an SOP, the agency is responsible for all aspects of enforcement within the state. A state agency that becomes an accredited certifying agent serves the same function—certifying organic production, processing, and handling—as any other certifying agency would. In the list below, you can see how three states are working in this area.

- **California.** The California Department of Food and Agriculture (CDFA) administers an SOP only and does *not* certify.

- **Washington.** The Washington State Department of Agriculture is an accredited certification agency only, not an SOP.

- **Utah.** The Utah Department of Agriculture and Food administers an SOP and is also an accredited certifying agent.

In order to qualify as an SOP, a state must adopt its own set of regulations or laws that accept the standards set by the NOP. Upon approval by USDA, those state laws essentially become the NOP laws for that state. COPA (the current California law), for example, incorporates by reference the federal regulations and is interpreted in conjunction with them, and it is written so that any future changes to the federal law automatically become regulations for

California. Compliance and enforcement provisions are part of both the NOP and approved SOPs. Most states do not have SOPs; they fall under the direct jurisdiction of the federal NOP with regard to enforcement activities.

NOP IMPLEMENTATION IN CALIFORNIA

California's SOP is administered through CDFA, responsible for fresh agricultural commodities (e.g., fruits, nuts, vegetables, field crops), meat, poultry, and dairy products, and the California Department of Public Health (CDPH), responsible for processed food products, cosmetics, and pet food. The SOP's purpose is to protect producers, handlers, processors, retailers, and consumers of organic foods sold in California by enforcing labeling laws related to organic claims for agricultural products. Its activities are coordinated with the California Organic Products Advisory Committee, USDA, and the California county agricultural commissioners.

The California SOP's responsibilities include administration of the program, training county staff, investigating complaints, registering producers and certification organizations, and acting as a resource for information on COPA and the organic industry in California. The SOP is funded entirely by producer and certifier registration fees, a portion of which are used to support county enforcement activities. Because COPA brings two separate statutes relating to fresh and processed products together into one law, it is important that each individual carefully read the sections that relate to his or her type of business operation. The complete text of COPA is available online at the CDFA Organic Program Web site, www.cdfa.ca.gov/is/i_&_c/organic.html. If you are involved in both aspects—growing and processing—you should note the specific requirements pertaining to each.

Producers

Organic producers and handlers (e.g., wholesale distributors, retailers, and some processors who market direct to retail or consumers) are regulated through CDFA. California requires that organic producers be registered and, depending on their sales revenue, certified as well. All organic growers, regardless of gross sales, must be registered with CDFA. In addition, any producer who grosses more than $5,000 per year in total sales of organic products must be certified through an accredited certification agency.

Registration, which involves providing a map of the production area, a list of crops that you intend to produce as organic, and a 3-year history of substances or materials applied, is handled by your local county agricultural commissioner. You have to provide verification of the land use history and pay an initial minimum registration fee of $75, higher if gross sales are or are expected to be higher than $25,000 annually. After initial registration, fees for yearly renewal follow the schedule in table 1.1.

Certification follows a separate process. NOP regulations establish who must become certified before they can use the "organic" label on their products. The following general guidelines apply:

- **Exempt from having to be certified:** Producers who are grossing less than $5,000 per year in retail sales. Product must be produced, processed, and packaged under the producer's own label; product cannot be sold to anyone else for processing or packaging, although under the producer's own label the product can be marketed in retail stores.

- **Excluded from requirement of having to be certified:** Any operation or company that does not

repackage or relabel the product (e.g., bulk wholesale distributor) or retail establishments that process or prepare organic product on the same premises at which they sell to the customer.

- **Must be certified:** All other operations involved in production, processing, and marketing.

If you are not sure whether you need to be certified or what category you fall into, contact the CDFA SOP office at (916) 445-2180 (online at www.cdfa.ca.gov/is/i_&_c/organic.html). The steps involved in certification are described in more detail under "Certification Agents," later in this chapter.

Processors

As defined in COPA, every person in California who processes, packages, stores, distributes, or handles food, pet food, or cosmetics in California that are sold as organic (with the exception of processed meat, fowl, or dairy products) is required to register with the California Department of Public Health. Registration is handled through CDPH's Organic Processed Product Registration Program, and processors must pay an annual registration fee. Like producers, they must be both registered and certified.

CDPH itself handles registration of organic processors. To get an application, call CDPH or visit the CDPH Web site, www.cdph.ca.gov/programs/Pages/FDB.aspx. The application must be filled out completely and returned to CDPH with the appropriate fee. Upon receipt of the application and fee, the operation is registered, provided that if the operation processes food for human consumption it already has a valid processed food registration permit issued by CDPH. There is no preinspection or verification requirement, but during the operation's regular food safety inspection, the inspector may ask questions about the organic aspects of the operation. The registration fee is set according to the schedule shown in table 1.2. Fees for yearly renewal follow the same schedule.

It is possible that an operation or business will have to register both as producer and processor with both agencies (CDFA and CDPH). For example, olive growers who also make and sell olive oil must be registered with both CDFA and CDPH. The registration fee for CDFA is based on the market value of the olives prior to processing, whereas the fee for CDPH is based on the market value (gross sales) of the finished olive oil.

Table 1.1. Fee schedule for companies required to be registered with the California Department of Food and Agriculture (CDFA)

Gross sales	Annual registration fee
$0–4,999	$25
$5,000–10,000	$50
$10,001–25,000	$75
$25,001–50,000	$100
$50,001–100,000	$175
$100,001–250,000	$300
$250,001–500,000	$450
$500,001–1,000,000	$750
$1,000,001–2,500,000	$1,000
$2,500,001–5,000,000	$1,500
$5,000,001–15,000,000	$2,000
$15,000,001–25,000,000	$2,500
$25,000,001 and above	$3,000

Note: Fees are subject to change; check with CDFA for the most current fee schedule. Producers that sell processed product pay fees based on the value of raw product prior to its being processed and the value of any product sold as unprocessed. Other exceptions are outlined in the text of COPA (see the CDFA Organic Program Web site, www.cdfa.ca.gov/is/i_&_c/organic.html).

Table 1.2. Fee schedule for companies required to be registered with the California Department of Public Health (CDPH)

Gross annual sales or revenue	Annual registration fee
$0–5,000	$50
$5,001–50,000	$100
$50,001–125,000	$200
$125,001–250,000	$300
$250,001–500,000	$400
$500,001–1,500,000	$500
$1,500,001–2,500,000	$600
$2,500,001 and above	$700

Note: Fees are subject to change; check with the CDPH or CDFA for current fee schedule.

Certification, which is distinct from registration, involves the same steps and requirements for processors as it does for growers (see the section "Certification Agents" later in this chapter). The same restrictions and exemptions described above for producers also apply to processors. If your farming business is involved in both production and processing, it makes sense for you to find a certifying agency that is accredited to certify both aspects (production and processing) of the operation. Most certifiers do both, but it is wise to make sure. Some certifiers consider postharvest handling to be a separate activity requiring a separate application, fee, and inspections. Other certifiers see processing as an addendum or attachment to the production side of the operation, an approach that can reduce costs and paperwork for the grower. This discretion on the part of the certifier only goes just so far: Any activities beyond simple postharvest handling (e.g., beyond storage or transportation to market) would clearly fall under the processing rules and require separate registration and certification for each component.

If you are not sure whether you need to be certified or what category applies to your operation, contact the CDPH Food Safety Program at (916) 650-6500. The remainder of this chapter covers the certification process and is written mainly for growers, but the general principles and steps involved in becoming certified apply to processors as well.

Certification Agents

An organic producer can choose from any of the certification agencies accredited by USDA and should select the one that will best serve his or her needs and budget (see sidebar, "How to Evaluate and Select a Certifier"). As of spring 2008, more than 90 accredited certification agents were approved by the USDA, including state and international agencies as well as private companies. About 20 of those were registered under the California SOP. All certifiers who want to operate in California must register with CDFA. Most certifiers require an application fee as well as inspection fees. These are substantial, starting at around $500. If you intend to export your operation's organic products outside of the United States, you will also need certification from the certifying body in the destination country or from the International Federation of Organic Agricultural Movements (IFOAM). Some certifiers that are accredited through the NOP have direct partnerships with those foreign certifiers, meaning that they are accredited by those international agents and can verify international certification in addition to NOP certification.

The certifying agency monitors grower practices to ensure that the grower is in compliance with NOP regulations. The certifier

- must know and enforce the NOP standards

- must verify the producer's compliance with national standards through annual scheduled inspections of the client's records, fields, and production and handling areas

- may perform annual soil and tissue tests from the client's operation in accordance with NOP guidelines

- may perform a certain number of surprise inspections every year

- must conduct ongoing review and inspection of client operations to ensure compliance

As a third-party agent of the NOP, the certifier is forbidden to provide advice on the producer's operation but can and must provide information to the producer about the certification process and the legal requirements for maintaining certification. This means that the certifier should know the client's operation in detail and should have current knowledge of NOP regulations and be able to communicate them effectively but may not act as a consultant to the grower. Certification agents

HOW TO EVALUATE AND SELECT A CERTIFIER

Questions to ask yourself before you begin to contact certifying agents

1. Will my product be sold only within California? If so, you may want to contact only agents with home offices in California. If the product will be sold outside California, other certifiers, including those based in other states, can be considered.

2. Will my product be sold outside the United States? If not, you only need NOP certification. If you are exporting to other countries, you should look for agents that certify according to international standards, such as the Japanese Agricultural Standard (JAS), the European Union (EU), or the International Federation of Organic Agriculture Movements (IFOAM), and who have knowledge of those markets.

3. Is my product a raw crop or a processed food? Which agents seem to be strongest and provide the best service in relation to my final product? Do some research on the certifiers you are considering.

4. Who certifies my friends and other companies like mine in the county or state? Talk to those friends and companies and find out how well they are satisfied with their certifying agent.

Questions to ask certifying agents after you have checked the NOP Web site and confirmed that the certifying agent is accredited by the NOP

1. Given the products, land, or processing that I propose to certify, what are the expected costs for the initial submission of a certification request, actual certification on the first inspection, and expected ongoing annual costs? When are the fees due and payable?

2. How long have you been certifying organic operations?

3. Do you offer training programs in any aspect of organic compliance?

4. How do you determine which materials are allowed and which are prohibited?

5. Is your cost or fee structure linked to organic sales, overall sales, number of acres, or complexity of the operation?

6. Once paid, are any of my fees refundable if I decide to withdraw from application or from certification? Are there any additional costs or hidden fees?

7. Are there any discounts for renewal clients?

8. How many pages would a typical application package be for an operation similar to mine?

9. I am going to export to a foreign country [give name of country or markets]. What is your expertise in certifying for this target market? What documents or transaction forms are required for export of organic products to this country? Are there any fees for generation of specific documentation required for each individual shipment?

10. What are your policies regarding confidentiality of information and records?

11. What are the charges if I need to add or change something (land, crop, product, SKU of same product, etc.) during the year and what documentation is required?

12. Will you assist in any manner in marketing organic commodities?

13. Do you have a Web site listing the companies you have certified? Do you list companies you have certified by commodity?

14. Do you certify co-handlers under my application? In that situation, is the fee reduced?

15. What type of verification is needed to prove land history?

Other considerations

All certifiers are private companies or are a part of a government entity. Each has a different ability to respond to you and to provide services. Check the level of service by finding out:

1. What is the average time that it takes the certification agency to complete a certification?

2. What is the average time that it takes the agent to call back or respond to a question you may have?

3. Do they provide supporting information that helps you understand compliance issues?

4. What kind of support does the certifier offer for helping you understand and complete paperwork?

5. Does the certifier provide a detailed explanation outlining the sections of the NOP that back up their certification positions and decisions?

6. Are the inspectors friendly and courteous? Are the inspectors employees of the certifying agency or do they contract with independent inspectors? Do you have the option of choosing your inspector?

Notes

The NOP's official list of accredited certification agencies is published on its Web site at www.ams.usda.gov/nop. The *New Farm's Guide to U.S. Organic Certifiers* is a very useful tool for learning about and comparing accredited certification agencies in the United States. For more information visit the New Farm Web site, http://newfarm. rodaleinstitute.org/ocdbt/.

must maintain their accreditation through the NOP and, as noted in the sidebar, may also become certification agents for foreign or international bodies. Some certifiers may also serve a broader role in organic trade associations or in dealing with policy issues.

The Organic System Plan and Growers' Responsibilities

Growers should recognize that they also have serious responsibilities and commitments in relation to certification. According to NOP regulations, every certified organic producer is required to develop and keep current a production or handling organic system plan (OSP) that describes the operation and the practices and procedures to be maintained and also provides a list of each substance to be used as a production or handling input. The OSP includes

- land history

- sources of all seed and planting stocks used by the organic operation

- all inputs to the crop and all materials applied to the crop

- list of all practices and procedures for soil and pest management, fertility, and crop nutrition that are used in the operation, including details on monitoring practices

- explanation of barriers or buffers used to prevent commingling between organic and nonorganic products or drift from nonorganic operations

- harvest and postharvest practices, including equipment use

- a description of records kept to prove compliance

- any additional information needed to document NOP compliance

Most certifiers provide a form that growers can use to prepare their OSP. These forms may be helpful for some growers, but they are not required; it is possible to write your own OSP, so long as you include all the required information. In either case, writing the OSP is useful for planning your farming operations for the year. Some certifiers may provide help in preparation of the OSP. The length of time to complete an OSP depends on the level of complexity of the farming operation.

The OSP is central to the certification process. It serves as a management tool to help farmers make decisions and react to changing circumstances. It also describes the human and natural resources of a farm, helps a producer manage those resources in an integrated way, and can help the grower with budgeting and financial planning. Last and most important, the OSP constitutes a legally binding contract between the certifier and the certified operation. Breach of that contract can result in denial or loss of certification. This last point is crucial, and growers must understand that the records they keep constitute the only proof that their contract (the OSP) has been fulfilled. The OSP is the commitment or promise to the certifier that production and handling will be carried out in a certain way, but the grower must show through good record keeping how he or she has kept that promise. Growers must get approval from their certifier if they are going to deviate from their submitted OSP.

In summary, the federal Organic Foods Production Act (OFPA) is a performance regulation, not a prescriptive regulation. This means producers decide how they will grow or process their products and how they will demonstrate compliance with the NOP. The OSP is the individual grower's or operator's specific plan on how he or she is going to meet those performance targets. Growers should be proactive and create the road map for achieving compliance that works best for them.

TRANSITION AND STEPS FOR ORGANIC CERTIFICATION

Transition Period

In order to be certified organic, a farming site cannot have had any prohibited substances applied to it for 36 months. One way to accomplish this is to choose a site that has not been farmed for at least the last 3 years. If the site has been actively farmed, the grower must ensure (and verify with accurate records) that no synthetic, noncompliant materials were used during the 3-year period. The 3-year transition period is a time when growers can educate themselves about the NOP and certification requirements (see the list of Internet resources at the end of this chapter). This is also an opportunity to begin work on a soil-building program, an integral component of the OSP. A grower in transition may also consider hiring a consultant to help prepare the farm for certification. This is not required and adds to the transition's cost, but a good consultant could help the grower avoid mistakes that might otherwise slow the farm's progress toward certification. Upon obtaining certification, the grower must implement all production practices addressed in the regulation and must be proactive in building up soil organic matter with compost, cover crops, or other amendments, using organically acceptable methods to control weeds, insects, diseases, and any other pests, and implementing biodiversity. The grower must document in detail the name and amount of every material applied to each field and must keep the OSP current. A producer does not need to be certified or registered during the transition period.

Keeping Records from the Start

It is very important that growers keep careful records of the exact date they began the organic transition, the last prohibited materials applied to the field and when they were applied, and all of the specific inputs and practices used during the 3-year transition period. When you apply for certification, the certifier will need detailed production practice records in all of those areas. We recommend that you keep receipts and labels of organic inputs and that you get verification from neighbors, pest control advisers (PCAs), county agricultural commissioner's pesticide use reports, or other local officials to corroborate that you ceased using prohibited pesticides (and when you ceased) and when you started the organic transition. Also, it is helpful to keep records of production activities or tasks by recording work orders or maintaining other labor records.

Contacting the Certifier

The certifier will send you all of the application materials you will need to get started. Certifiers expect growers to read the handbooks or literature they provide. The grower's OSP must be fully prepared and submitted with the appropriate forms. Many certifiers require that growers sign an affidavit that shows land use history and affirms the truthfulness of the application. It usually takes 2 to 5 months for the application to be reviewed. Certifiers charge certification fees, which generally include

- a one-time application fee (ranging from $100 to $300)

- inspection costs, which usually pay for an inspector's time to travel to and inspect the farm and prepare a report, usually once per year

- annual fees (some certifiers use a fee based on acreage; others charge an annual fee of some set percentage of the gross production value)

Make sure you know the specific fees that the certifier will charge before you apply. The application fee must be paid when you submit the application. The fees cover the costs of the certification procedures and agency overhead and pay for the time of the inspection staff. Through certification, growers gain transparent third-party verification of their compliance with NOP. This does open the farming operation to greater public scrutiny, but ultimately it enhances the value of the farm's products in the marketplace.

Review of Application, Inspection, and Notification of Certification

The certifier will review the grower's application and inspect the operation before certification is approved. First, a trained inspector calls the grower to set up the initial inspection. After the inspection, the inspector submits a report to the certifier for review. The certifier informs the grower of the certification status or lets the grower know about any further requirements the grower must meet to achieve or maintain certification.

During an inspection, the inspector will want to review or see records about

- land use history

- pesticide use reports

- cleaning or clean-out procedures if you use shared equipment

- your understanding of the regulations (e.g., you have read the appropriate manuals)

- the use of approved materials only

- appropriate systems to ensure compliance

- product labels for products or inputs that you use

Annual Review and Inspection

Organic certification inspections are done on farms once a year as required by the NOP. The inspector generally informs the grower of each visit in advance and sets up an appointment for the inspection. Certifiers and state organic inspectors do, however, have the right to make unannounced visits or inspections as well. If an inspector encounters violations or problems, the grower will receive a notice of minor noncompliance or major noncompliance from the certifier. If the noncompliance is minor, the grower is reprimanded, told not to repeat the error, and may be sent a reminder about the particular issue. If the inspector considers the noncompliance to be major, the grower's certification status may be affected. The grower may be required to submit requested documents or update a section of the OSP, or in serious cases the grower can be decertified.

As discussed previously, certification inspectors are not allowed to provide advice or information to the grower while conducting an inspection. They do not serve as consultants or advisors. This is intended to ensure that the certifier's role is one of enforcement and compliance checking. However, some certification agencies do have departments or divisions that provide information or educational materials to all growers, separate from the certification enforcement. In addition, many organizations, services, advisors, Web sites, and other resources provide information about organic practices.

A summary of the certification process and the interaction between the grower and the certifier is outlined in table 1.3.

Issues Regarding Compliant and Noncompliant Materials

All materials that can be used in crop and livestock production are classified into one of four categories under the NOP. Note that in the context of the NOP regulation, the words nonsynthetic and natural are used synonymously.

- **Category 1: Allowed nonsynthetic.** Most nonsynthetic (natural) materials are allowed, with the exception of those that are specifically prohibited (category 2, below). There is no list of allowed natural materials; the rule of thumb is that natural materials are allowed unless they are specifically prohibited.

- **Category 2: Prohibited nonsynthetic.** A few nonsynthetic (natural) materials, such as arsenic and strychnine, are not allowed under the NOP. See NOP Rule sections 205.602 and 205.604.

- **Category 3: Allowed synthetic.** These synthetics are allowed, but their use is restricted to specific purposes as defined in the NOP regulation and National List. Use of these materials for other purposes is prohibited. See NOP Rule sections 205.601 and 205.603.

Table 1.3. Roles and responsibilities of growers and certifiers

Grower	Certifier
Transition (3-year period)	
Read and understand National Organic Program (NOP) regulations. Gather information. Do research and planning. Begin transition.	
Document NOP-compliant practices and management for 3 years. Select an accredited certification agent.	The certifier would have a role only if grower were to seek transitional certification.
Request application materials and instructions from certifier.	Provide materials and information to client (e.g., NOP regulations, application materials, etc.).
After 3 years	
Register with CDFA prior to sale of organic products.	Certifier is not involved in registration; grower registers through CDFA.
Submit application and organic system plan (OSP) to certifier for certification. (This should be done well in advance of the first anticipated organic harvest [6 months to 1 year ahead of time].)	Verify that the grower is registered with CDFA and review the application for certification in a timely manner. Request additional information from grower as necessary.
Supply any new information or details requested by the certifier to demonstrate compliance. Grower must submit to on-site inspection.	Conduct inspection and notify grower of the results. Upon approval, the OSP becomes a legally binding contract.
Implement the OSP. Record practices and procedures. Obtain prior approval for any alterations to the OSP. Keep accurate and detailed records. Keep records organized in preparation for annual inspection.	Annually review and inspect the grower's operation to ensure continuing compliance for certification.

- **Category 4: Prohibited synthetic.** Most synthetic materials are not allowed. The exceptions are those specifically approved for use in organic production (category 3, above).

The National List of allowed materials for organic production is overseen and maintained by the NOP (www.ams.usda.gov/nop/NationalList/ListHome.html). The National List indicates generic compounds and materials that are allowed, but does not list specific brand name products. Brand name information is available through the Washington State Department of Agriculture (http://agr.wa.gov/Portals/Org) and the Organic Materials Review Institute (OMRI) (www.omri.org). OMRI is a private nonprofit organization that provides verification and listings of products that meet the national organic standards. It publishes and regularly updates two lists: a brand name list and a generic list.

As a certified grower, you are expected to keep track of updates and changes in the approval status of the materials that you use. Before trying new materials, you should research the NOP, OMRI, and WSDA lists to make sure that the materials are allowed, and also get permission from your certifier to amend your OSP and use the product. Do not rely solely on the verbal or written declarations of vendors. Consult a reliable source to confirm any vendor claims.

The certification agency has the final decision on the acceptability of inputs for each farm operation. In making a determination about the acceptability of an input, the certification agency must evaluate the input based both on its ingredients and on the context in which it will be applied. Prior to using any input on an organic farm, you must obtain written approval (e.g., a certificate of compliance) from your certifier. The certifier will contact you or the material's manufacturer if he or she needs any additional information to determine whether use of the material is in compliance. A word of caution: Any input that you use in your farming operation without the prior written approval of the certifier could be viewed as a departure from your agreed-upon OSP and could be grounds for adverse action on the part of the certifier, USDA, or state programs.

REFERENCES/RESOURCES

Dimitri, C., and C. Greene. 2002. Recent growth patterns in the U.S. organic foods market. USDA Economic Research Service Publication AIB-777. www.ers.usda.gov/Publications/aib777.

Economic Research Service, USDA. 2011. Organic agriculture: Organic market review. www.ers.usda.gov/briefing/organic/demand.htm.

Klonsky, K. 2010. A look at California's organic agriculture production. Giannini Foundation of Agricultural Ecomonics, UC Davis. http://agecon.ucdavis.edu/extension/update/articles/v14n2_3.pdf.

Klonsky, K., and K. Richter. 2005. A statistical picture of California's organic agriculture: 1998–2003. Davis: University of California Agricultural Issues Center. http://aic.ucdavis.edu/oa/pubs.html.

Klonsky, K., and L. Tourte. 1998. Statistical review of California's organic agriculture: 1992–1995. Davis: University of California Agricultural Issues Center, Statistical Brief No. 6 (May). http://aic.ucdavis.edu/research1/organic.html.

———. 2002. Statistical review of California's organic agriculture: 1995–1998. Oakland: University of California Agriculture and Natural Resources, Publication 3425. http://aic.ucdavis.edu/research1/organic.html.

OTA (Organic Trade Association). 2010. US organic industry overview. www.ota.com/pics/documents/2010OrganicIndustrySurveySummary.pdf.

INTERNET RESOURCES

ATTRA (National Sustainable Agriculture Information Service), http://attra.ncat.org/organic.html.

California Organic Program, www.cdfa.ca.gov/is/i_&_c/organic.html.

eOrganic, www.eorganic.info.

Organic Farming Compliance Online Handbook, www.sarep.ucdavis.edu/organic/complianceguide.

Organic Materials Review Institute, www.omri.org.

Scientific Congress on Organic Agricultural Research (SCOAR), www.ofrf.org/networks/scoar.html.

Washington State Department of Agriculture Organic Food Program, http://agr.wa.gov/Portals/Org.

USDA National Organic Program, www.ams.usda.gov/nop.

Business and Marketing Plans for Organic Operations: An Overview

LAURA TOURTE AND MARK GASKELL

A business plan and a marketing plan are essential, foundational tools for a farm business. These plans can be structured as individual documents or combined to form one comprehensive report. Business and marketing plans have multiple functions that include but are not limited to

- defining and describing a business

- identifying vision, goals, and objectives

- formulating strategies and techniques for conducting a business

- determining what resources are on hand and what additional resources are needed

- assessing the risks and challenges associated with operating a business and marketing its products

- seeking partners, financing from a lending institution, or venture capital to begin or continue a business

Finally, these plans can be used as powerful tools to monitor, evaluate, and guide a business over time, modify decisions and operations when necessary and appropriate, and plan for future contingencies.

This chapter looks at the content and major sections of business and marketing plans, especially as they apply to organic agricultural operations. Additional print and online resources that are available to help you develop your own plans are also included in this chapter.

PLAN CONTENTS AND DEVELOPMENT

It takes creativity, research, coordination, and organization to develop business and marketing plans. It is important to note, however, that there is no single format or right way to construct these plans. Some plans are succinct and concise, while others are lengthy and elaborate. Content, organization, and clarity are noteworthy characteristics of superior business and marketing plans. Ultimately, each plan should be tailored so that it represents the needs of the business and contains only as much information as is necessary to fully and accurately portray the operation. Following are a brief description of the main sections of business and marketing plans, and, where appropriate, selected considerations that particularly apply to organic enterprises (table 2.1). A plan should include these sections:

- Cover page

- Executive summary

- Table of contents

- Business description

- Marketplace analysis

- Products and services

- Sales and marketing

- Management

- Financial information

- Appendixes

Table 2.1. Section content of a combined business and market plan

Section	Content
Cover page	Includes all pertinent business information, including owner name(s) and contact information.
Executive summary	Presents a brief overview of the major aspects of the entire plan. It should be constructed as a stand-alone document.
Table of contents	Lists all major plan sections, including page numbers. Subsections may also be listed.
Business description	Presents a concise, declarative statement describing the business, identifies products and services, sources of funds, and competitive edge.
Marketplace analysis	Presents background research data and information on emerging trends and their link to the business.
Products and services	Describes products and services in more detail, including potential new product development.
Sales and marketing	Identifies and describes the target market and its link to products and services, the business's strengths and weaknesses, and projected income and profits.
Management	Describes administrative structure and other key personnel, plus skills and experience of each.
Financial information	Presents historical data as well as 1 to 3 years' projected budgets, including cash flow. Risk and contingency plans can also be included here.
Appendixes	Presents support documents, including résumés and graphics.

Cover page. A plan's cover page must include all pertinent business information, such as the name of the business, the date the plan was written, the names and titles of business owners, contact information, Web site listings, and any recognizable labels or logos for the business.

Organic considerations. Include any distinguishing organic identifiers; for example, the logo of the operation's accredited certifying agent, programs in which the operation participates, or other seals as allowed (e.g., USDA Organic). These identifiers serve as important market tools.

Executive summary. This section summarizes key points from all other sections of the plan. Because the executive summary is likely to be the first—and possibly only—section that on-the-go business people or potential investors will read, it is important that it be sufficiently complete to work as a stand-alone document. The executive summary is essentially a shorter, more concise version of the complete plan. It must contain the major points and elements of the plan, but it must also be short—one to three pages in length. This section is best written after all other parts of a business plan have been

clearly developed. One example is provided in the sidebar on the facing page.

Table of contents. A table of contents lists and identifies all major plan sections that follow, with page numbers included. You can also identify subsections within major sections here.

Business description. A business description contains a clear, concise statement of what the business is as well as its products and services. For example, "Coastal Organic Farm is a vegetable and small fruit farming operation that produces and markets high-quality fresh and specialty organic products direct to consumers through farmers markets and to small independent retail operations in the Central Coast region of California." This statement is simple, declarative, and sufficient for the purposes of this section.

The business description further includes a mission statement, or overarching vision for the business, short-term and long-term goals and objectives, information on how the business is or will be funded, resources that are currently available, and a brief explanation of what makes the business unique or competitive with respect to other similar businesses.

SAMPLE EXECUTIVE SUMMARY FOR COASTAL ORGANIC FARM

Background and Goals

Coastal Organic Farm is a proposed small to mid-scale organic operation (gross annual sales under $500,000) that, for its size, expects to become a strong and reliable competitor in the production and (self-) marketing of high-quality fresh and specialty vegetables and small fruits. Its primary function is to ensure consistency in the diversity, volume, and quality of organically produced mixed vegetables and berries. The operation consists of 35 acres, a portion of which will be in production each year. It will house office space, a packing facility, and equipment sufficient to coordinate the transportation of product to market. Coastal Organic Farm is targeted for start-up because of its location in one of the premier agricultural regions in the state as well as its proximity to large urban areas (San Francisco, San Jose) with a demonstrated demand for diverse, high-quality, specialty ethnic, and organic products.

Marketplace Analysis

California leads the nation in production and value of fresh fruits and vegetables, with roughly 50 percent of the United States total. Statistics show that between 1976 and 2009, per-capita consumption of fresh and processed fruits and vegetables increased by 11 percent. More specifically, demand for specialty ethnic, vegetable, and organic products shows potential for notable growth in the future. Coastal Organic Farm recognizes that organic growers of various scales (large, medium, and small) already conduct business in the Central Coast area. However, market research also shows that opportunities exist for competitive, business-oriented operations with diverse, high-quality products that are locally or regionally produced. Coastal Organic Farm does not intend to compete with the larger players mentioned above, but rather to carve out a niche for its size and scale of production.

Products and Services

Coastal Organic Farm will function as a producer and marketer of high-quality fresh and specialty agricultural products. As such, the operation will produce mainstream crops with demonstrated demand (e.g., lettuce and broccoli) and specialty crops that show increasing demand (e.g., chili peppers and Asian vegetables).

Sales and Marketing

Coastal Organic Farm will take a proactive, professional business-oriented stance with respect to sales and marketing. It will market primarily to California-based wholesalers, as well as to local and regional independent retail grocers. Coastal Organic Farm has developed a relationship with, and received a commitment from, organic produce wholesalers and a group of coastal natural foods grocers for California and locally grown specialty ethnic vegetables, herbs, and small fruits. In addition to the business orientation mentioned above, Coastal Organic Farm will attract and retain buyers of its products by keeping abreast of new and emerging market trends and research, with special emphasis on changes in demographics and consumer demand. This information will be evaluated periodically and incorporated into the farm's market strategies. Please also refer to the larger sales and marketing section of this plan for greater detail and analysis.

Management and General Operations

Coastal Organic Farm recognizes that top-notch management is critical to the success of an organization. Equally important are building strong business relationships and honoring commitments. During the start-up period, the owners will serve as managers and marketers to handle daily operations for the farm business, set priorities, manage finances, plan for contingencies, and plan activities for the future. The owners expect to recruit and hire a professional business manager/marketer as the operation expands.

Organic considerations. This section should also contain a reference to the growers' or partners' experience with organic agriculture in addition to their wider experience in agricultural enterprises; for example, the total number of years the operation has been in business, when (or if) it transitioned from conventional to organic management, and the number of years it has functioned as an organic operation. The business section should further reference participation in or compliance with state registration and federal organic certification programs. You can also include information that highlights the characteristics that differentiate an organic operation from a conventional business. For example, discuss any special structures or equipment that are necessary to store and handle bulky production materials, or other facilities and systems that may be necessary to maintain organic product integrity as required by state and federal regulations.

Marketplace analysis. The marketplace analysis compiles background research data with information on the marketplace's current status and emerging trends. This information can include prices and shipping volumes, statistics on demographics and consumer trends, market competition, and expected profit margins. Social and environmental issues that influence the marketplace might also be considered. It can also include information about the supporting infrastructure, such as the availability of suppliers, cooling and shipping facilities, or other transportation and handling resources.

Sources of information include (but are not limited to) local, state, and federal government agencies, such as the USDA Agricultural Marketing Service (USDA–AMS), university Cooperative Extension offices, trade associations, trade publications, other support organizations, other farmers, and industry suppliers. This information should be as current as possible, it should be realistic, and it should support a rationale, connection, or need for the business.

The Internet can also provide helpful information. The USDA–AMS Market News Web site, for example, is an especially valuable resource because it provides historical and current wholesale prices and shipping volumes, arranged by shipping point source, for a wide range of products in diverse major domestic and foreign cities. It also has historical and current data on shipping point volumes from diverse U.S. and foreign shipping points, arranged by product type.

Organic considerations. Research data and reporting systems are not as well developed for organic products and markets as they are for agriculture as a whole. This is especially true with respect to historical data and information. Many information resources, however, have recently made significant strides in tracking and compiling background data, statistics on organic agriculture, and information on consumer preferences and trends. For example, the USDA–AMS recently made available web-based reports online for organic shipping volumes, point sources, and prices. A resource list specific to organic agricultural information is included in table 2.2.

Products and services. The next section is essentially an expanded description of the business's products and services. Products that will be the

mainstays of an operation should be identified and briefly discussed. For example, crop type(s), number of producing acres, and expected yield and quality could be included here. Information on services that the business may provide should also be offered here, such as custom farm work or cooling and shipping for neighboring operations.

A discussion of plans for new products and their development, as well as anticipated challenges and pitfalls and how they might be dealt with, also belong in this section. Unique aspects of particular products and services—those that fill a particular niche or that can be differentiated in the marketplace—should be included here.

Organic considerations. A business or marketing plan for an organic operation should note any special requirements or needs for the business and how these requirements will be addressed. For example, describe the availability and ease of sourcing organic inputs and access to farm services, cooling, and shipping facilities that cater to organic operations. If you are proposing new products or changes to an existing business, it is important that you include potential successes and challenges and tell how these would be managed within the context of the organic industry.

Sales and marketing. The sales and marketing section of a farm business plan is arguably the most important. It encompasses not only the act of selling products or services, but also the identification, development, and retention of customers. Consumers drive the farm marketplace, signaling their preferences to producers and buyers through their purchasing patterns and habits. It is important that farm owners and managers carefully evaluate the marketplace and then strategically target their preferred market segments. An effort should be made to project anticipated yields—in different size classes if appropriate—along with anticipated unit costs and sales prices during distinct production periods. You can then consolidate these figures into overall average sales prices and unit costs based on your anticipated volumes (also see the "Financial Information" section, below). In other words, it is vital that the farm business identify and select one or more potentially valuable market segments, understand consumers within each segment, and anticipate and plan the products and services accordingly.

Table 2.2. Selected informational resources for organic enterprises

Agency/organization	Web site (URL)
ATTRA National Sustainable Agriculture Information Service	www.attra.org
CDFA California Organic Program	www.cdfa.ca.gov/is/i_&_c/organic.html
Organic Farming Research Foundation (OFRF)	ofrf.org
Organic Materials Review Institute (OMRI)	www.omri.com
Organic Trade Association (OTA)	www.ota.org
The Rodale Institute Organic Price Report (OPR)	www.rodaleinstitute.org/Organic-Price-Report
UC Davis Agricultural Issues Center	aic.ucdavis.edu/research1/organic.html
UC Davis Current Cost & Return Studies	coststudies.ucdavis.edu
UC Farm Business & Market Place	ucce.ucdavis.edu/farmbusinessandmarketplace
UC Small Farm Program	www.sfp.ucdavis.edu
UC Sustainable Agriculture Research & Education Program (SAREP)	www.sarep.ucdavis.edu/Organic/
UC Vegetable Research & Information Center (VRIC)	vric.ucdavis.edu/veg_info_topic/organic_production.html
USDA Agricultural Marketing Service & Market News (for price and volume data)	www.ams.usda.gov marketnews.usda.gov
USDA Economic Research Service (ERS)	www.ers.usda.gov/
USDA National Organic Program (NOP)	www.ams.usda.gov/nop/
U.S. Small Business Administration (for guidance in writing a business plan)	www.sba.gov

Sales and marketing plans often include measurable objectives along with strategies designed to meet those objectives. For example, one objective might be to improve profit margins on a subset of offered products by 2 percent over the next year. A strategy to meet this objective might be to evaluate current price and cost structures, project different price and cost scenarios, and determine whether the objective is realistic and attainable. The plan would then describe the steps or techniques that would be used to implement such a strategy or to adjust objectives periodically as experience suggests new, more efficient alternatives.

As a part of this section, farm owners should also assess their strengths and weaknesses with respect to selling and marketing. This may result in the hiring of a person with specialized organic sales or marketing experience if the farmer is lacking in these skills. Finally, farmers should project their potential income from sales as well as the costs associated with both production and marketing.

Organic considerations. An organic product, by law, requires special handling and record keeping to ensure the product's integrity as it moves through market channels. An organic-focused business or market plan might therefore include in its sales and marketing section a description of the operational strategies and procedures used to maintain product integrity. If appropriate, make note of potential pitfalls (such as a lack of experienced sales representatives or wholesalers), how the business proposes to work through such constraints, and anticipated impacts on the business. Information on historical trends and prices for organic products is limited. However, research-based information with respect to costs of production and potential returns to growers is becoming more and more available. Table 2.2 lists current sources of information for organic cost and price structures.

Management. The management section includes a description of the administrative structure of the business, including owners, managers, and other key personnel. The responsibilities of each person should be briefly discussed along with the individual's previous farm or business experience or other relevant skills and experience. Résumés and other background details may be included as supplemental documentation in the appendixes section if desired.

Organic considerations. Points to include in this section include any personnel's experience with special relevance to organic production, postharvest handling, labeling, food safety, sales, and marketing, and their level of familiarity with or understanding of organic regulations and the organic industry as a whole.

Financial information. Not surprisingly, the financial information section of a business plan is essential. This part of the plan could include any historical production and price data that a business has available. For new businesses, this information might come from university Cooperative Extension research, local agricultural commissioner crop reports, or USDA statistics. Continuing businesses are likely to have this information documented in their farm records. One to three years of detailed projected business budgets, including cash flow, are also important elements of this section. Income statements and balance sheets may also be included here, or they may be included in the appendixes.

Every farm business has a degree of risk associated with its operations. In the financial information section, potential sources of risk should be identified and discussed. Examples include production risks, such as crop failures from natural disasters, and market risks, such as unpredictable price fluctuations. Risk also arises when production is pushed forward or back into periods with short supply but higher prices, whether due to a planned

extension of the production season or to poor climatic conditions. Even with its risks, pushing a significant proportion of the production into target market windows can still be a sound business strategy. Strategies to handle or minimize risks—for example, crop diversification or crop insurance—should also be discussed here.

Some business plans include contingency plans in the financial information section. Contingency plans demonstrate that a farmer understands the potential for unexpected change and has alternative tactics or strategies on hand to cope with these changes. Contingency planning can take the form of simple, declarative "if…, then" statements or more detailed scenario plans.

Organic considerations. As previously discussed, you can use resources listed in table 2.2 to assist with the information needs of an organic enterprise. For established organic operations, required record keeping provides an excellent source of information from which to compile financial information. Organic enterprises typically have crop rotation and diversification plans as a part of their organic farm plan. A reference to or summary of these plans could be included in the financial information section to demonstrate the business's prudent preparation for risk.

Appendixes. The appendixes include supporting documents or other information relevant to the proposed or continuing business. These documents can include additional financial documents such as balance sheets and income statements, a business history, résumés, and graphics.

INTEGRATION INTO BUSINESS

Once constructed, business and marketing plans become important working documents or tools to guide and direct a business. Owners can refer to the plans periodically to change the business vision, monitor and evaluate the achievement of goals and objectives, use identified strategies to meet objectives or overcome challenges, and modify farm operations and business decisions when appropriate.

REFERENCES/RESOURCES

Many resources are available to farmers to help them develop business and marketing plans; several are listed in table 2.2 and here. Some of these resources are interactive in that they present questions to which the reader can respond as well as detailed worksheets that can help in the development and construction of plans.

DiGiacomo, G., R. King, and D. Nordquist. 2003. Building a sustainable business: A guide to developing a business plan for farms and rural businesses. Saint Paul, MN: Minnesota Institute for Sustainable Agriculture. Beltsville, MD: Sustainable Agriculture Network. www.sare.org/Learning-Center/Books/Building-a-Sustainable-Business.

Grubinger, V. P. 1999. Sustainable vegetable production from start-up to market. Ithaca, NY: Cornell University Cooperative Extension. Natural Resource, Agriculture, and Engineering Service.

Pisoni, M. E., and G. B. White. 2002. Writing a business plan: A guide for small premium wineries. Ithaca, NY: Cornell University Department of Applied Economics and Management, EB 2002–06. http://hortmgt.aem.cornell.edu/resources/publications.htm.

Richards, S. 2002. Getting started in farming? Five keys to success. Ithaca, NY: Cornell University Department of Applied Economics and Management, Smart Marketing Series.

Schlough, C. 2001. Knowing your market—The most challenging part of a business plan. Ithaca, NY: Cornell University Department of Applied Economics and Management, Smart Marketing Series.

Uva, W. L. 1999. Travel the road to success with a marketing plan. Ithaca, NY: Cornell University Department of Agricultural, Resource, and Managerial Economics, Smart Marketing Series. http://hortmgt.aem.cornell.edu/resources/publications.htm.

Western Extension Marketing Committee. 2008. Niche markets: Assessment and strategy development for agriculture. Tucson: University of Arizona Department of Agricultural and Resource Economics, www.valueaddedag.org.

White, G. B., and W. L. Uva. 2000. Developing a strategic marketing plan for horticultural firms. Ithaca, NY: Cornell University Department of Agricultural, Resource, and Managerial Economics, EB 2000–01. http://hortmgt.aem.cornell.edu/resources/publications.htm.

Economic Performance of Organic Vegetables: The Case of Leaf Lettuce on the Central Coast

LAURA TOURTE, RICHARD F. SMITH, KAREN KLONSKY, AND RICHARD L. DE MOURA

ASSESSING COSTS OF ORGANIC PRODUCTION

Conversion of a farm to certified organic techniques can involve cost and risk. Nonetheless, organic vegetables can be profitable in a well-managed fresh market operation. Price premiums for organic vegetables are an important stimulus to conversion and certification, but it is difficult to predict the stability of premiums over time given variability in market supply and consumer demand. For 2009, sales of California-grown organic products were $921 million, almost twice the $510 million sold in 2005 (Klonsky and Richter 2010). For 2005—the year for which the most recent organic vegetable statistics are available—organic vegetable sales were estimated to be $355 million, or 39 percent of the total dollar value for all organic sales. The Central Coast posted the highest organic vegetable sales, followed by the San Joaquin Valley and the South Coast. Together, these three regions account for 85 percent of all organic vegetable production in the state. From 2005 to 2009, organic vegetable sales in California increased by 38 percent.

Net returns for organic vegetables fluctuate with production and market conditions. It is therefore important that organic vegetable production decisions be made with full knowledge of local technical and marketing options. If reliable organic pest, soil, and field management systems can be devised for a given site, the grower needs to develop and implement a transition management plan that should be continually evaluated, updated, and compared to conventional practices, costs, and yields.

A number of basic economic tools can be used to help you analyze business decisions for an organic vegetable operation. These include whole farm budgets, enterprise budgets, partial budgets, and capital budgets. The type of budget you use in an analysis depends on your needs as a vegetable grower or farm manager. The whole farm budget is appropriate when an entire business is to be evaluated. Enterprise budgets help determine a specific commodity's potential profitability. Partial budgets enable growers and managers to examine the financial impact of specific changes within a cropping system; for example, changing pest management techniques or practices. Capital budgets are used to assess the impact of long-term business changes such as the purchase of land or equipment.

AN EXAMPLE: ESTIMATED COSTS AND EXPECTED RETURNS FOR CENTRAL COAST ORGANIC LEAF LETTUCE

An enterprise budget for 2009, including a monthly cash flow analysis, for a double-cropped 5-acre organic leaf lettuce operation on the Central Coast (Santa Cruz and Monterey Counties) is presented in tables 3.1 to 3.3. Costs are based on representative cultural practices; these practices are not used every year or by all growers. Estimated cost tables are intended to help you understand costs, make production decisions, determine potential returns, prepare your own budget, and evaluate production loans. Each farming situation varies.

Following are the underlying assumptions used to derive the cost and return estimates:

• **Farm size.** The total farm size in our example is 200 acres. Organic leaf lettuce is planted on five acres; the remaining 195 acres are planted to other organically grown vegetables such as broccoli, cauliflower, and celery, and possibly small fruits such as strawberries or raspberries. The farmer rents the land at $2,200 per acre per year. Because the land can produce up to two cash crops per acre per year, only part ($1,200) of the annual rent is charged in this example. As of this writing, the cost is within a realistic range of rental values for row crop land on the Central Coast.

- **Planting.** Leaf lettuce is direct seeded (organic pelleted) into 40-inch, double-line beds using a precision planter. No particular lettuce variety is specified for this study.

- **Yield.** Yield for organic leaf lettuce varies depending on season and growing conditions. Yields range from 500 to 1,000 24-count, 25-pound boxes per acre. The assumed yield for this study is 750 boxes per acre, which falls within the range of yields reached by growers in this area.

- **Labor.** Basic hourly wages for workers in the example are $13.00 and $10.00 for machine operators and fieldworkers, respectively. Adding 35 percent for payroll taxes, workers' compensation, social security, and other benefits increases the labor rates to $17.55 per hour for machine labor and $13.50 per hour for field labor. The labor hours for operations that involve machinery are 20 percent higher than actual operation time due to the extra time involved in equipment setup, moving, maintenance, work breaks, and repair.

- **Interest on operating capital.** The interest rate is based on cash cultural and harvest costs. It is calculated monthly until harvest at the rate of 5.75 percent per year. This interest can be viewed as the cost of a production loan. If your business is self-financed, you can interpret the interest as the opportunity cost of tying up capital in the crop.

Tables 3.1, 3.2, and 3.3 all provide information for the hypothetical organic leaf lettuce operation described in the set of assumptions. Each table presents the corresponding enterprise budget information in a different format and, as such, contains different levels of detail on production practices, inputs, and costs.

Table 3.1 details the material and labor costs associated with the operations that appear in table 3.2. The quantities, costs per unit, and costs per acre of each input are presented. Table 3.2 lists the cash and labor costs by cultural operation. The operations are listed in order of occurrence and the costs of the material inputs are listed. Cash and noncash overhead costs are included to calculate the total costs of production, excluding management and risk (to avoid redundancy, table 3.1 does not categorize the cash and noncash overhead costs). Table 3.3 lists the total operating costs by operation from table 3.2 on a monthly basis. The total cash costs per month are included.

For the hypothetical organic leaf lettuce operation presented, the total cash costs are estimated to be $8,994 per acre at a yield of 750 25-pound boxes per acre. When you include capital recovery on buildings and equipment, the total costs are estimated at $9,112 per acre.

For product marketed as organic, total revenue per acre (gross returns) is calculated by multiplying the yield (number of boxes per acre) by the price per box. The price to growers is expected to range from $10 to $19 per 25-pound box. Net returns represent revenue to growers minus a specific set of costs; for example, operating costs or cash costs. Net returns above total costs are considered profit, or returns to management and labor. One useful measure is to calculate the break-even price and yield for the operation. For the break-even price,

Table 3.1. 2009 material costs ($/acre) for organic leaf lettuce production on the Central Coast of California*

Costs	Quantity per acre	Unit	Cost per unit ($)	Cost per acre ($)
Operating Costs				
Fertilizer/soil amendments				
Compost: manure/green waste (haul/spread) (½ cost [tonnage] to lettuce)	2.50	ton	40.00	100
Gypsum (½ cost [tonnage] to lettuce)	.50	ton	42.00	21
Pelleted chicken manure	1,000.00	lb	0.25	250
13-0-0 bloodmeal	450.00	lb	0.75	338
6-1-1 Phytamin 801	171.00	lb	0.51	87
Seed				
Cover crop (cereal-legume mix) (¼ cost of seed to lettuce)	30.00	lb	1.20	36
Leaf lettuce (organic)	148.20	thousand	1.00	148
Alyssum	0.05	lb	15.00	1
Irrigation				
Water: pumped	17.00	ac-in	8.33	142
Drip tape (10 mil) (½ cost [footage] to lettuce)	6,541.00	foot	0.03	196
Insecticide				
Dipel DF	1.00	lb	15.99	16
Pyganic 1.4 EC	2.00	pint	24.87	50
Contract/custom				
Ground application (insects)	1.00	acre	25.00	25
Harvest (box, pick, haul, supervision)	750.00	box	4.15	3,113
Harvest (palletize, cool)	750.00	box	1.10	825
Sell commission, 8% of $15	750.00	box	1.20	900
Pest management consultant	1	acre	30.00	30
Spread green waste + gypsum mixture	3	acre	10.00	30
List and fertilize	1	acre	25.00	25
Labor (machine)	8.61	hr	17.55	151
Labor (nonmachine)	43.28	hr	13.50	584
Fuel, gas	5.95	gal	3.36	20
Fuel, diesel	63.06	gal	3.70	233
Lube				38
Machinery repair				47
Interest on operating capital @ 5.75%				80
Total operating costs/acre				**7,485**

* Some totals may appear to be inaccurate due to rounding.

Table 3.2. 2009 costs for organic leaf lettuce production on the Central Coast of California*

Operation	Operation time (hr/acre)	Labor ($/acre)	Fuel & repairs ($/acre)	Materials ($/acre)	Custom/rent ($/acre)	Total ($/acre)
Cultural costs						
Fertilize: preplant (gypsum/compost) (½ cost to lettuce)†	0.00	—	—	121	30	151
Land prep: subsoil (½ cost to lettuce)†	0.61	13	51	—	—	64
Land prep: disk & roll 2X (½ cost to lettuce)†	0.29	6	25	—	—	31
Land prep: chisel 2X (½ cost to lettuce)†	0.35	7	29	—	—	37
Land prep: land plane field 2X (½ cost to lettuce)†	0.24	5	21	—	—	26
Cover crop: plant 1X every 2 yrs (¼ cost to lettuce)†	0.04	1	2	36	—	39
Cover crop: mow 1X every 2 yrs (¼ cost to lettuce)†	0.04	1	2	—	—	3
Cover crop: disk 2X every 2 yrs (¼ cost to lettuce)†	0.07	2	6	—	—	8
Land prep: disk & roll 1X	0.14	3	13	—	—	16
Land prep: list beds; fertilize (pelleted chicken manure)	0.00	—	—	250	25	275
Irrigate: pre-irrigate, sprinkle	2.00	27	—	17	—	44
Land prep: cultivate 2X (rolling cultivator)	0.21	5	9	—	—	14
Land prep: shape beds & roll	0.23	5	10	—	—	15
Plant: lettuce	0.28	9	13	148	—	169
Insect: plant insectary (Alyssum seed)	0.07	1	2	1	—	4
Irrigate: sprinkle 3X	3.00	41	—	25	—	65
Stand establishment: thin & weed (hand)	16.25	219	—	—	—	219
Weed: cultivate	0.11	2	4	—	—	7
Irrigate: lay drip line & laterals (drip tape)	1.00	63	43	196	—	301
Fertilize: sidedress 1X (bloodmeal)	0.20	4	5	338	—	347
Irrigate: drip 5X	0.75	10	—	100	—	110
Fertilize: through drip (Phytamin)	0.00	—	—	87	—	87
Pest: worms (Dipel) & aphids (Pyganic)	0.00	—	—	66	25	91
Weed: cultivate & furrow 2X (break bottoms)	0.21	5	9	—	—	13
Weed: hand-hoe	12.00	162	—	—	—	162
Irrigate: retrieve drip & laterals	1.50	113	62	—	—	175
Pest: pest management consultant	0.00	—	—	—	30	30
Pickup use	1.43	30	26	—	—	56
Total cultural costs	**41.02**	**732**	**331**	**1,384**	**110**	**2,557**
Harvest costs						
Cut, pack, haul	0.00	—	—	—	3,113	3,113
Cool, palletize, sell	0.00	—	—	—	1,725	1,725
Total harvest costs	**0.00**	**—**	**—**	**—**	**4,838**	**4,838**

Operation	Operation time (hr/acre)	Labor ($/acre)	Fuel & repairs ($/acre)	Materials ($/acre)	Custom/rent ($/acre)	Total ($/acre)
Postharvest costs						
Chop stubble	0	3	7	—	—	11
Total postharvest costs	0	3	7	—	—	11
Interest on operating capital @ 5.75%		—	—	—	—	80
Total operating costs/acre		735	338	1,384	4,948	7,485
Cash overhead costs						
Land rent						1,200
Office expenses						127
Field sanitation						63
Liability insurance						2
Annual organic certification fees						90
Property taxes						7
Property insurance						6
Investment repairs						12
Total cash overhead costs						1,509
Total cash costs/acre‡						8,994

Investment	$/Per producing acre	Annual cost capital recovery ($/producing acre)	Total cost
Noncash overhead costs			
Building	400	25	25
Shop tools	75	6	6
Fuel tanks	23	1	1
Pipe: sprinkler	66	8	8
Trailer, pipe #1	11	2	2
Trailer, pipe #2	11	2	2
Equipment	695	74	74
Total noncash overhead costs	1,280	117	117
Total costs/acre			9,112

Notes:

* Some totals may appear to be inaccurate due to rounding.

† Costs of land preparation done prior to first planting that affected the lettuce and second crop on same land are split equally between the lettuce and second crop.

‡ Some growers prefer to separate harvest costs from total cash costs to reflect growing costs: total cash costs minus harvest costs equals total growing costs. For this table, that would be $8,994 – $4,838 = $4,156.

Table 3.3. 2009 monthly cash costs ($/acre) for organic leaf lettuce production on the Central Coast of California*

Costs	Oct 08	Nov 08	Dec 08	Jan 09	Feb 09	Mar 09
Cultural costs						
Fertilize: gypsum/compost (½ cost to lettuce)	151	—	—	—	—	—
Land prep: subsoil (½ cost to lettuce)	64	—	—	—	—	—
Land prep: disk & roll 2X (½ cost to lettuce)	31	—	—	—	—	—
Land prep: chisel 2X (½ cost to lettuce)	37	—	—	—	—	—
Land prep: land plane field 2X (½ cost to lettuce)	26	—	—	—	—	—
Cover crop: plant 1X every 2 yrs (¼ cost to lettuce)	39	—	—	—	—	—
Cover crop: mow 1X every 2 yrs (¼ cost to lettuce)	—	—	—	—	—	3
Cover crop disk 2X every 2 yrs (¼ cost to lettuce)	—	—	—	—	—	8
Land prep: disk & roll 1X	—	—	—	—	—	—
Land prep: list beds; fertilize (pelleted chicken manure)	—	—	—	—	—	—
Irrigate: pre-irrigate, sprinkle	—	—	—	—	—	—
Land prep: cultivate 2X (rolling cultivator)	—	—	—	—	—	—
Land prep: shape beds & roll	—	—	—	—	—	—
Plant: lettuce	—	—	—	—	—	—
Insect: plant insectary (Alyssum seed)	—	—	—	—	—	—
Irrigate: sprinkle 3X	—	—	—	—	—	—
Stand establishment: thin & weed (hand)	—	—	—	—	—	—
Weed: cultivate	—	—	—	—	—	—
Irrigate: install drip line & laterals (drip tape)	—	—	—	—	—	—
Fertilize: sidedress 1X (bloodmeal)	—	—	—	—	—	—
Irrigate: drip 5X	—	—	—	—	—	—
Fertilize: through drip (Phytamin)	—	—	—	—	—	—
Pest: worms (Dipel) & aphids (Pyganic)	—	—	—	—	—	—
Weed: cultivate & furrow 2X (break bottoms)	—	—	—	—	—	—
Weed: hand-hoe	—	—	—	—	—	—
Irrigate: retrieve drip & laterals	—	—	—	—	—	—
Pest: pest management consultant	—	—	—	—	—	—
Pickup use	6	—	—	—	—	6
Total cultural costs	**353**	**—**	**—**	**—**	**—**	**16**
Harvest costs						
Cut, pack, haul	—	—	—	—	—	—
Cool, palletize, sell	—	—	—	—	—	—
Total harvest costs	**—**	**—**	**—**	**—**	**—**	**—**
Postharvest costs						
Chop stubble	—	—	—	—	—	—
Total postharvest costs	**—**	**—**	**—**	**—**	**—**	**—**
Interest on operating capital 5.75%	**2**	**2**	**2**	**2**	**2**	**2**
Total operating costs/acre	**354**	**2**	**2**	**2**	**2**	**18**
Cash overhead costs						
Land rent	—	—	—	—	—	—
Office expense	25	—	—	—	—	—
Field sanitation	—	—	—	—	—	—
Liability insurance	—	—	—	—	—	—
Annual organic certification fees	—	—	—	—	—	—
Property taxes	7	—	—	—	—	—
Property insurance	6	—	—	—	—	—
Investment repairs	1	—	—	—	—	1
Total cash overhead costs	**40**	**—**	**—**	**—**	**—**	**1**
Total cash costs/acre	**394**	**2**	**2**	**2**	**2**	**19**

* *Note:* Some totals may appear to be inaccurate due to rounding.

Apr 09	May 09	Jun 09	Jul 09	Aug 09	Sept 09	Oct 09	Nov 09	TOTAL ($/acre)
–	–	–	–	–	–	–	–	151
–	–	–	–	–	–	–	–	64
–	–	–	–	–	–	–	–	31
–	–	–	–	–	–	–	–	37
–	–	–	–	–	–	–	–	26
–	–	–	–	–	–	–	–	39
–	–	–	–	–	–	–	–	3
–	–	–	–	–	–	–	–	8
–	–	–	–	16	–	–	–	16
–	–	–	–	275	–	–	–	275
–	–	–	–	44	–	–	–	44
–	–	–	–	14	–	–	–	14
–	–	–	–	15	–	–	–	15
–	–	–	–	169	–	–	–	169
–	–	–	–	4	–	–	–	4
–	–	–	–	44	22	–	–	65
–	–	–	–	–	219	–	–	219
–	–	–	–	–	7	–	–	7
–	–	–	–	–	301	–	–	301
–	–	–	–	–	347	–	–	347
–	–	–	–	–	27	56	27	110
–	–	–	–	–	29	58	–	88
–	–	–	–	–	91	–	–	91
–	–	–	–	–	7	7	–	13
–	–	–	–	–	–	162	–	162
–	–	–	–	–	–	–	175	175
–	–	–	–	8	8	8	8	30
–	–	–	–	11	11	11	11	56
–	–	–	–	598	1,068	302	221	2,557
–	–	–	–	–	–	–	3,113	3,113
–	–	–	–	–	–	–	1,725	1,725
–	–	–	–	–	–	–	4,838	4,838
–	–	–	–	–	–	–	11	11
–	–	–	–	–	–	–	11	11
2	2	2	2	5	10	11	35	80
2	2	2	2	603	1,078	313	5,104	7,485
–	–	–	–	–	1,200	–	–	1,200
–	–	–	–	25	25	25	25	127
–	–	–	–	–	–	31	31	63
–	–	–	–	–	–	–	2	2
–	–	–	–	–	–	–	90	90
–	–	–	–	–	–	–	–	7
–	–	–	–	–	–	–	–	6
–	–	–	–	2	2	2	2	12
–	–	–	–	28	1,230	59	151	1,509
2	2	2	2	630	2,308	372	5,255	8,994

this entails selecting an expected yield and then calculating the price you would have to receive in order to make your revenue equal your costs at that expected yield. Similarly, calculating the break-even yield requires that you select an expected price and then calculate the yield at which total income would equal the costs of production. For the farming operation studied here, which has an assumed yield of 750 boxes per acre and prices at or above $12.15 per box, organic leaf lettuce is a profitable crop. At a midpoint price of $14.50 per 25-pound box, the break-even yield would be 531 boxes per acre. Actual economic performance depends on numerous factors, including variety, site conditions, yearly production conditions, consumer demand, and market competition.

Cost of production studies for many crops, including organically grown crops, are available from the University of California Cooperative Extension.

Visit us online at http://coststudies.ucdavis.edu or contact your local UC Cooperative Extension office. Studies are updated periodically and analyses of additional crops are performed each year.

REFERENCES/RESOURCES

Klonsky, K., and K. Richter. 2010. Statistical review of California's organic agriculture, 2005–2009. Davis: University of California Agricultural Issues Center, http://aic.ucdavis.edu/research1/organic.html.

Tourte, L., R. F. Smith, K. M. Klonsky, and R. L. De Moura. 2009. Sample costs to produce organic leaf lettuce, double-cropped. Davis: University of California Cooperative Extension, Central Coast Region (Monterey and Santa Cruz Counties), http:/coststudies.ucdavis.edu/files/lettuceleaforganiccc09.pdf.

PHOTO: KATHY KEATLEY GARVEY

PHOTO: KATHY KEATLEY GARVEY

Soil Fertility Management for Organic Crops

MARK GASKELL, RICHARD F. SMITH, JEFF MITCHELL, STEVEN T. KOIKE, CALVIN FOUCHE, TIM HARTZ, WILLIAM HORWATH, AND LOUISE JACKSON

ROLE OF ORGANIC MATTER AND HUMUS

Increasing the organic matter content of soil is a key aspect of an organic crop production system. The formation and decomposition of soil organic matter are fundamental life-promoting processes that store and release the energy derived from photosynthesis. Soil organic matter is mainly the product of the microbial and faunal decomposition of plant residues. Decomposition of plant residue leads to the formation of humic substances, which constitute 70 to 80 percent of the organic matter in most soils. The remaining soil organic matter, termed light fraction or particulate organic matter, is composed of a continuum of material ranging from recently deposited litter to highly decomposed materials that are no longer recognizable as plant residues. Soils with a higher clay content in temperate climates generally have the greatest soil organic matter content. In California, organic matter typically makes up no more than 1 to 3 percent of soil dry weight in cultivated agricultural soils and 4 to 6 percent in untilled pasture soils. Studies have shown that normally it is not possible to increase soil organic matter by more than 1 percent, but even a relatively small increase can dramatically improve the soil's fertility.

During the formation of soil organic matter, nutrients such as nitrogen (N), phosphorus (P), and sulfur (S) are incorporated into the soil structure, allowing it to act as a reservoir of these and other nutrients. The decomposition of soil organic matter releases nutrients and they become available for plant uptake. Generally, 2 to 5 percent of soil organic matter decomposes every year. Soil organic matter consists of a number of fractions that vary in their composition and activity. Humus is the most resistant and mature fraction of soil organic matter. It is very slow to decompose and may last for hundreds of years. High carbon (C) and low nitrogen residues such as straw or corn stalks decompose slowly but are efficient producers of humus. High nitrogen plant residues such as young cereals and legumes decompose quickly, producing less humus. Although the process of organic matter formation is not well understood, it is clear that humus improves soil properties and crop growth and that practices should be employed to increase soil organic matter.

The decomposition of organic matter in soils can provide significant quantities of several important nutrients. A portion of the N from organic amendments is converted into plant-available mineral forms such as ammonium (NH_4^+) and nitrate (NO_3^-) through a process called mineralization; however, the timing and amount of mineralization are often out of sync with the crop's need, making in-season fertilizer applications necessary. This disjunction between the timing of N mineralization from organic matter and the crop's N uptake is a major challenge in organic systems. Organic matter is a good source of P as well. As P is mineralized from organic matter it either becomes available for plant growth or is bound to soil minerals. Organic matter is also a significant source of S and micronutrients such as iron, copper, and zinc.

Besides supplying nutrients, soil organic matter also influences soil fertility by enhancing the soil's chemical and physical attributes. Soil organic matter can bind nutrients through the process of cation exchange. Ammonium, calcium (Ca), magnesium (Mg), and potassium (K) are nutrient ions that are held on cation-exchange sites on organic matter. The cation-exchange capacity of soil organic matter can contribute from 20 to 70 percent of the soil's total cation-exchange capacity.

Soil structure is influenced by the association of soil organic matter with minerals to form aggregates. Aggregate formation improves the soil's structure and its water infiltration and increases its water-holding capacity. These changes improve root growth and provide habitat for a diversity of soil organisms.

HOW TO DETERMINE NUTRIENT NEEDS

Together, a crop's nutrient requirements and the soil's nutrient-supplying capacity dictate the management practices necessary for successful crop production in a given location. Soil testing is essential for the assessment of nutrient levels and is often a requirement for organic certification. Management of nutrients such as P, K, Ca, Mg, and S should be directed toward raising these nutrients to optimum levels in the soil, as determined by a soil test. The availability of P in acidic to mildly alkaline soils (pH > 6.0) is assessed by means of the Olsen bicarbonate test; for more acid soils (pH < 6.0), the Bray test is used. Most vegetable production areas in California have soil pH values greater than 6.0, so our discussion here will focus on the Olsen soil test. Naturally occurring levels of P in most California soils were initially below 30 ppm. After years of fertilization for commercial vegetable production, fields now routinely have soil P content in excess of 60 ppm along the coast, with somewhat lower values in the interior valleys. The availability of P in the soil is reduced in low soil temperatures (< 60°F) and, as a result, crops grown in the cooler part of the year may need more P applications for good growth. Approximate values for P in the soil (from the bicarbonate P test) for warm- and cool-season vegetables are listed below:

- Warm-season vegetables: 20–25 ppm P

- Cool-season vegetables: 50–60 ppm P

Compost and some organic fertilizers are good sources of P. It is important to monitor soil P levels on a yearly basis, as it is possible for it to rapidly build up to high levels when you use composts and other organic amendments. Excessive soil P can result in field runoff with high P concentrations, which can impair the quality of surface waters such as rivers, creeks, and lakes.

The best way to determine the K level in a soil is the ammonium acetate extraction test. In general, if the soil K level is greater than 200 ppm, additional applications are unlikely to increase yield. However, maintenance applications may help replace the K that the crop removes from the soil. For soils with a K level below 150 ppm, fertilizer application is warranted. Composts and some organic fertilizers are good sources of K.

Ca, Mg, and S are usually present in sufficient quantities in the soil and in irrigation water to adequately supply a crop's needs. In very sandy, low-organic-matter soils, S availability may be limited, but normal organic practices (e.g., compost application, organic S fertilizer, use of S as a fungicide) will typically keep an adequate amount of S in the soil. While neither Ca nor Mg availability is often limiting for crop nutrition, relatively low soil Ca or high Mg content can result in poor soil structure and slow water infiltration in some fields. In such circumstances, the application of gypsum (naturally occurring calcium sulfate) is the most appropriate remedy.

In an organic system, you cannot accurately determine that your N management is appropriate based on a simple soil test. Unlike conventional production, where N management is based on the use of soluble, readily available N fertilizers, N management is based on manipulating organic sources of N. Organic N must be mineralized through the action of soil microbes before it becomes available for plant uptake. This process can supply a significant quantity of N over time, but any estimation of the amount and timing of N mineralization is very complicated since so many factors affect the process. The most important of those factors are soil temperature (below 50°F, mineralization is insignificant, but above that temperature mineralization increases as soil temperature increases), soil moisture (mineralization proceeds rapidly in moist soils but is inhibited by either excessively wet or dry conditions), and tillage practices (soil tillage stimulates a temporary burst of microbial activity that then declines over the course of days or weeks). Despite the complicated interactions of these factors, you can make a rough estimate of mineralization from soil organic matter based on the amount of organic N present in the soil and the percentage of that amount that is likely to mineralize over a given period of time. The following procedure describes a method for estimating the timing for this process.

The first step is to estimate how much organic N is in the soil. You can do this directly with a specialized laboratory test or infer it from the soil's organic matter content. In most agricultural soils, organic N constitutes approximately 7 percent of soil organic matter. The vast majority of N mineralization takes place in the top 1 foot of the soil. Using a standard estimate of soil weight, (4,000,000 lb dry soil per acre-foot), the organic N content of a soil with 1 percent organic matter would be

4,000,000 lb soil ×
0.01 (i.e., 1% organic matter) ×
0.07 (i.e., 7% N in organic matter) =
2,800 lb organic N/acre

The second step would be to estimate the percentage of soil organic N that is likely to mineralize during the crop cycle. Laboratory incubation studies of dozens of California soils have shown that, at best, about 2 percent of soil organic N is mineralized in a 2-month period at 77°F. For a soil with 1 percent organic matter, that would be

2,800 lb organic N/acre ×
0.02 (i.e., the 2% of organic N that mineralizes) =
56 lb plant-available N/acre

The 2 percent estimate for N availability for a short-term annual crop can be adjusted to fit field-specific conditions based on the factors previously described. Fields that are sprinkler irrigated will keep the entire soil surface moist, while much of the surface soil in drip irrigated fields may be very dry. Soil temperature for spring and fall crops will be lower than for summer crops. Fields in which the grower practices any form of reduced tillage will tend to have slower N mineralization. Heavy clay soils are more readily waterlogged by rain or irrigation, and effective N mineralization may be reduced. It is important to note that this technique for estimating N mineralization from soil organic matter does not take into account the N contribution from recently incorporated crop residue, compost, or other organic amendments. Those contributions are described elsewhere in this chapter.

Synchronizing N mineralization from soil organic matter, cover crop residues, and organic amendments to maintain adequate N availability for crop production is a challenge. The generalized pattern of N mineralization and crop N uptake is presented in figure 4.1. The rate of N mineralization from soil organic matter and recently incorporated residues and amendments typically peaks before the crop reaches its maximum N uptake rate. Even in organic systems, N loss through leaching or denitrification

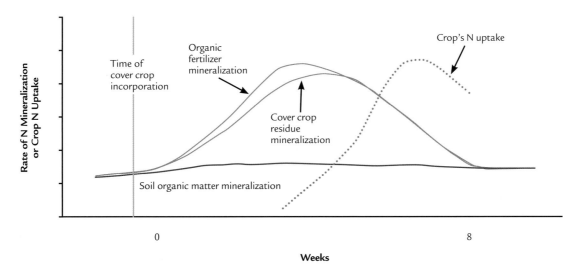

Figure 4.1. Timing of N mineralization from soil organic matter, cover crop residue, and organic fertilizer in relation to the crop's N uptake.

Table 4.1. Nitrogen requirement of vegetable crops based on seasonal N uptake

Low total N content (<120 lb/acre)	Medium total N content (120–200 lb/acre)	High total N content (>200 lb/acre)
baby greens	carrot	broccoli
beans	corn, sweet	cabbage
cucumber	garlic	cauliflower
radish	lettuce	celery
spinach	melon	potato
squashes	onion	
	peppers	
	tomato	

(conversion of nitrate to gaseous N in wet soil, and subsequent loss to the atmosphere) can be substantial if excessive water (from rain or irrigation) is applied to the field in the early weeks of the growing season.

Short-season crops with low N requirements, such as leafy greens and radishes (table 4.1), may produce well with the N available from soil organic matter plus cover crop residue or a compost application. Crops with a higher N requirement and longer growing season often need supplemental sidedress applications of organic N fertilizer. For many vegetable crops, quality is as important as yield: Product size, color, and uniformity can be critical, and N management is often the key to maximizing these attributes.

NUTRIENT SOURCES

Cover crops. Cover crops fix and trap nutrients, add organic matter to soils, and reduce nitrate leaching, nutrient runoff, and soil erosion. In California, cover crops are widely used in organic farming systems because the climate is mild enough to support growth during the fall, winter, and early spring in most crop production areas. Nonleguminous cover crops such as grasses and brassica species are preferred in situations where nutrient availability is high in the fall and where cover crops can trap nitrate and phosphate that would otherwise be lost to leaching or runoff. Nonlegumes also tend to be more tolerant of cooler temperatures than legumes. Legumes fix atmospheric nitrogen, at least when concentrations of mineral nitrogen in the soil are low, and thereby add to the net availability of

nitrogen in the cropping system. The mixture of legumes and grasses is a common strategy because the grass utilizes soil N and so prevents high soil N concentrations that might otherwise inhibit fixation. Mixtures also ensure that the overall cover crop is productive under a range of weather conditions.

In California, cover crops typically take up or fix between 100 and 200 pounds N per acre. Cover crops (e.g., legumes and younger stages of cereals and mustards) are often tilled into the soil when the carbon-to-nitrogen (C:N) ratio is less than 20 to 1 to achieve a net release of N to the soil in order to feed subsequent vegetable crops. High N content in cover crops reduces competition between the subsequent vegetable crop and the soil microbiota for mineral N. When cover crops with a low N content such as mature cereals (i.e., C:N ratio > 20) are incorporated into the soil, subsequent vegetable crops can be temporally N deficient because soil microbes use the available soil N to break down the cover crop residue. However, these higher C:N cover crops are instrumental in building soil organic matter, which in turn is advantageous for long-term soil fertility and improvements in soil physical properties. A longer-term grass or brassica cover crop is therefore recommended for intermittent use, as long as cropping patterns permit a sufficient crop-free period so the residue has time to decompose.

Less than half of the amount of N in a cover crop typically becomes available to the subsequent crop. Much of the cover crop N remains in resistant organic forms in soil organic matter, unavailable to plants. The organic N in the readily decomposable fraction of cover crop residues, however, can be

Table 4.2. Common organic fertilizer materials and nutrient analysis (%)

Material	Nitrogen (%)	Phosphorus (%)	Potassium (%)
Alfalfa meal	4	1	1
Blood meal	12	0	0
Bonemeal	2	5	0
Chicken manure, pelleted	2–4	1.5	1.5
Chilean nitrate	16	0	0
Feather meal	12	0	0
Fish meal or powder	10–11	6	2
Kelp	<1	0	4
Meat- and bonemeal	8	5	1
Potassium-magnesium sulfate	0	0	22
Processed liquid fish residues	4	2	2
Seabird and bat guano	9–12	3–8	1–2
Soft rock phosphate	0	16	0
Soybean meal	7	2	1

mineralized very rapidly into plant-available forms in the first few weeks after incorporation. The rate of mineralization of available N from a low C:N (< 20) cover crop increases over a 3- to 6-week period following incorporation and then returns to pre-incorporation levels by week 6 to 8 (see figure 4.1). A cover crop, then, can be a valuable source of short-term N, whereas longer-season vegetable crops following a cover crop rotation may require additional applications of N later in the season. When N from cover crops has been mineralized, it can be taken up and utilized by the crop or lost via leaching during spring rainy periods. For this reason, it is best to plant vegetable crops relatively soon after cover crop incorporation, although the cover crop should be allowed to decompose for at least 3 to 4 weeks in order to avoid potential stand establishment and pest problems that may otherwise arise as the residue decomposes.

Compost. Compost, particularly if it contains animal manure, can be a relatively cost-effective organic source of both macronutrients and micro-nutrients. When applying compost, the challenges are to know the composition of the compost and to understand how to use it most efficiently. The grower should understand the composting process that the supplier uses and should also know the sources of the compost's raw materials. If the materials that are being composted are low in nutrients,

the compost will have a low nutrient analysis. Poor-quality or immature compost may actually tie up nitrogen in the soil and decrease the availability of N to the growing crop. The C:N ratio of a compost is one indication of its N availability. As the C:N ratio rises above 20:1, the tendency increases for N from the soil to be tied up. A compost with a C:N ratio of less than 20:1 will generally release N to the succeeding crop. Other quality considerations are the age of the compost, particle size, pH, salt concentration, and purity (i.e., the volume of soil, sand, and other nonorganic materials mixed with the compost). The National Organic Program (section 205.203) describes additional standards on compost sources. Because compost analysis is based on dry weight, moisture content will add to the compost's weight and lower its nutrient analysis. It is not unusual for a commercial compost to have a moisture content of 25 to 30 percent.

Mineralization rates from compost application are relatively low, and compost application is usually a poor source of N to address an immediate need. Recent studies have shown that typically no more than 15 percent of the N in the compost is made available in the first year following incorporation. This may partly explain problems with N fertilization often observed during the transition from conventional to organic production. A longer-term program of repeated compost applications

Table 4.3. Net N mineralization (as a percentage of initial organic N) from organic fertilizers, as influenced by temperature and length of incubation

Product	Temp. (°F)	Percentage of initial organic N mineralized		
		After 1 week	After 4 weeks	After 8 weeks
Blood meal	59	41	60	64
	77	51	67	70
Fish powder	59	51	55	61
	77	48	60	64
Feather meal	59	42	56	59
	77	50	64	63
Poultry manure, pelleted	59	4	16	21
	77	10	23	36
Seabird guano	59	49	57	60
	77	45	48	54
Seabird guano, pelleted	59	42	61	64
	77	46	60	67

would be required to increase the overall amount of soil organic N and thereby increase the soil's N mineralization potential.

Manure. Aged animal manure can be a balanced source of N and other major and minor nutrients. Manure can only be used in organic production if it is applied to non-food-crop land, incorporated into the soil at least 120 days prior to harvest if the edible portion is in contact with the soil, or incorporated into the soil at least 90 days prior to harvest if the edible portion is not in contact with the soil. One potential limitation with using manure is the need for a consistent supply of a material that is sufficiently uniform so that you can confidently incorporate it into a production program. A public perception of food safety problems related to manure fertilization might further limit the use of manure. It is important to check with potential buyers to make sure they do not have internal policies restricting manure use. Most manure used in organic vegetable production is composted before field application, a step that minimizes food safety risks.

Commercial organic fertilizers. A number of approved organic fertilizers are available; some common examples are listed in table 4.2. Most of these materials are by-products of fish, livestock, and food processing industries. The commercial formulations and nutrient analyses of these materials vary considerably. In general, they range from 0 to 12 percent N and may also contain P or K. These fertilizers tend to be quite expensive, and in general their use is confined to situations where cover cropping or the application of compost is not feasible or has supplied insufficient available nutrients for the upcoming cash crop. The value of these fertilizers lies in the relatively rapid availability of their nutrients.

The short-term availability of nutrients depends largely on the nature of the fertilizer material and how it was processed. Table 4.3 compares mineralization rates for different organic N sources at different temperatures. These fertilizers can be applied prior to planting or in one or more supplemental side-dressings.

Materials such as processed liquid fish, liquid soybean meal, or sodium nitrate may also be applied through drip irrigation systems. Some of these products contain small particles that can be suspended in water but are not truly soluble. The water filtration necessary for micro irrigation systems may actually remove these particles, lowering the nutrient content that is ultimately delivered to the crop. Dilute liquid teas from these materials are sometimes applied to the soil or sprayed directly on the plant in an effort to improve nutrient availability, but the value of these teas as a nutrient source has not been clearly established. Chapter 5 discusses the use of compost teas as disease ncontrol agents.

Certain by-products of the meat processing industry, such as blood meal and bonemeal, have been restricted by some vegetable growers/shippers for market-related reasons. Mined Chilean nitrate (sodium nitrate), a source of rapidly available N, was an important component of organic fertilizer programs in the past because of its relatively low cost, solubility, and ease of use. In recent years, however, concerns that its mining and use present environmental risks, coupled with the widespread view within the organic movement that reliance on a soluble, mineral N fertilizer is incompatible with organic principles, have limited its use. The NOP currently restricts use of Chilean nitrate to no more than 20 percent of total N use and mandates that organic producers develop a plan to phase out the use of Chilean nitrate over time.

Minor element sources. Organic fertilizers commonly also contain one or more minor elements. A number of liquid materials and teas are also available that provide one or more minor elements. Some of these may be used in irrigation systems or applied to foliage. Field trials evaluating the effectiveness of minor element foliar applications when soil levels are already adequate do not show a consistent pattern of crop response. Costs for these materials vary widely. Synthetic fertilizers may be permitted by a certifying agency in specific circumstances for correction of minor element deficiencies. Some synthetic minor element fertilizers may contain contaminants such as hazardous materials that are used as fillers, and there is debate over whether the use of these materials, even in small amounts, is proper in organic production systems.

Special-purpose fertilizers. Specific, approved nutrient sources of K, Ca, and Mg may be useful to an organic grower when a soil test indicates some kind of deficiency or imbalance. Materials such as gypsum, lime, and potassium magnesium sulfate have been used in agriculture for many years. Their value is well proven, and organic growers can safely use them to correct deficiencies or imbalances of K, Ca, or Mg, or to raise soil pH. Gypsum is also

used to improve the water infiltration of soils with poor structure. Growers and researchers are still evaluating a number of other special-purpose fertilizers and growth enhancers. Materials derived from kelp and other processed seaweed contain nutrients and, in many cases, plant hormones and growth regulators. Some manufacturers claim that microbial soil stimulants enhance plant growth or reduce soil pests. Growers sometimes apply Brix mixes, humates, foliar nutrient and sugar solutions, and other materials as a way to raise nutrient or sugar levels (Brix) in plant sap in an attempt to improve the plants' resistance to pests. There is scant scientific data supporting the efficacy or cost effectiveness of such treatments.

CHARACTERISTICS OF ORGANIC FERTILIZERS

Organic fertilizer materials share a number of characteristics that distinguish them from conventional fertilizers. The key features to consider in a fertility management program are bulk, nutrient availability, and uniformity.

Bulk. Many organic fertilizer materials (e.g., composts or other organic by-products) are less-concentrated nutrient sources than are conventional fertilizers. Application rates for these materials are commonly 5 to 10 tons per acre, and sometimes more. An organic grower needs to consider how to transport, store, and apply such large quantities of material. Larger storage facilities and special handling equipment may be necessary.

Nutrient availability. Organic fertilizers often include a relatively small proportion of readily available soluble nutrients along with another nutrient fraction that either is permanently unavailable to the plant or will become available only gradually over time. As a general rule, these materials need to be applied earlier than conventional fertilizers in anticipation of plant nutritional requirements, often 2 to 4 weeks or more before the nutrients will be needed. The ultimate availability of nutrients will depend upon microbial activity, and poor mixing into the soil or extremes of soil moisture or temperature will delay nutrient availability. The composition and particle size of the material can also affect the rate of microbial decomposition and nutrient availability. A more concentrated, finer-textured material generally decomposes and releases its nutrients more readily than a coarser mixture.

Uniformity. Organic fertilizer materials can vary considerably with respect to particle size, moisture content, nutrient content, and nutrient distribution. Some of this inconsistency is inherent in the materials by virtue of the nature of the production process and the fact that organic materials often continue to change during transport and storage. Growers can get a better idea of the success of their efforts if they determine the chemical composition of key fertilizer materials and keep records of their variations over time. This may involve sending samples of composts and organic fertilizers to an independent laboratory for analysis.

The transitional period for a new organic operation can be the most demanding in terms of soil fertility management. This is because, though soil building and soil organic matter improvement are under way, their benefits have not yet been realized in a transitional field, and yet the grower is limited to the use of only a few soluble fertilizer materials. Growers gain experience during the transitional period and as the soil organic matter builds, its benefits are reflected in improved soil fertility. The USDA National Organic Program has developed lists of approved materials, which include a broad range of materials to provide for crop nutritional needs.

Also, be aware that materials from organic or natural sources may contain contaminants such as salts, heavy metals, and boron that may accumulate to toxic levels on a given field. Pick carefully from the materials that are available to you: Know the material's source and its composition.

REFERENCES/RESOURCES

Chaney, D. E., L. E. Drinkwater, and G. S. Pettygrove. 1993. Organic soil amendments and fertilizers. Oakland: University of California Division of Agriculture and Natural Resources, Publication 21505.

Gaskell, M., and R. F. Smith. 2007. Organic nitrogen sources for vegetable crops. HortTechnology 17:431–441.

Hartz, T. K., J. P. Mitchell, and C. Giannini. 2000. Nitrogen and carbon mineralization dynamics of manures and composts. Hortscience 35(2):209–212.

Miller, P. R., W. L. Graves, and W. A. Williams. 1989. Cover crops for California agriculture. Oakland: University of California Division of Agriculture and Natural Resources, Publication 21471.

Parnes, R. 1990. Fertile soil: A grower's guide to organic and inorganic fertilizers. Davis, CA: Fertile Ground Books.

Various authors. 1997–1999. Fundamentals of sustainable agriculture series. Fayetteville, AR: ATTRA (National Sustainable Agriculture Information Service), www.attra.org.

Various authors. n.d. Organic production: Recent publications and current information sources. Special Reference Brief SRB–01. Beltsville, MD: USDA, Alternative Farming Systems Information Center, National Agricultural Library, www.agnic.org.

Wyman, C., E. Chapman, and C. Sanders. 1990. Organic farming directory (revised). Oakland: University of California Division of Agriculture and Natural Resources, Publication 21479.

Managing Diseases for Organically Produced Vegetable Crops

STEVEN T. KOIKE, MARK GASKELL, CALVIN FOUCHE, RICHARD F. SMITH, JEFF MITCHELL, AND ALEXANDRA STONE

Plant diseases present challenging problems in commercial agriculture and pose real economic threats to both conventional and organic farming systems. Plant pathogens are difficult to manage for several reasons. First of all, the pathogens are hard to identify because they are microscopic. Positive identification of a pathogen often requires specialized equipment and training, and in some cases accurate diagnosis in the field is difficult.

Plant pathogens are constantly changing and mutating, re-emerging as new strains that present new obstacles to growers. Also, given today's local, regional, and international movement of seeds, plant material, and farming equipment, new and introduced pathogens periodically enter the California system, causing new disease problems.

The task of disease management is further complicated by the simultaneous presence of multiple types of pathogens. For any given crop, the grower may need to deal with a variety of fungi, bacteria, viruses, and nematodes. The situation is even more complicated for organic vegetable growers because they usually produce a wide array of vegetable crops and are prohibited from applying conventional synthetic fungicides. The world market is extremely competitive and continues to require that growers supply high-quality, disease-free produce with an acceptable shelf life. Disease management is therefore a critical consideration in organic vegetable production.

In an organic production system, it is most appropriate for a grower to develop disease control strategies that have an ecological basis. Insofar as possible, the organic system should encourage the growth and diversity of soil-inhabiting and epiphytic (plant surface-dwelling) microorganisms that have the potential to exert beneficial and pathogen-antagonistic influences. An increase in the genetic diversity of the crop host rotation is another management technique that encourages ecological diversity and health. The integration of disease management decisions with the overall decision-making process for insect and weed control and general production practices is also consistent with this approach.

RESISTANT PLANTS AND CULTIVARS

One of the most important components in an integrated disease control program is the selection and planting of cultivars that are resistant to particular pathogens (table 5.1). The term resistant, then, usually indicates the plant host's ability to suppress or retard the activity and progress of a pathogenic agent, and the resultant absence or reduction of symptoms. It is important, however, that we clearly establish a common understanding of the meaning of the term when discussing this quality with individuals from different sectors of the agricultural industry. Growers, researchers, plant breeders, and seed sellers may have slightly different definitions for the term. In addition, the word tolerant, which has a slightly different meaning, is used by some people interchangeably with resistant,

Table 5.1. Keys to using resistant cultivars and crops

- Constantly look for new and improved resistant cultivars.
- Get a clear, precise understanding of what *resistant* means with regard to the plants you purchase.
- Realize that plants' resistance factors may change depending upon the crop and pathogen.
- Budget for resistant cultivars (they may be significantly more expensive than nonresistant alternatives).

resulting in some confusion. By proper definition, tolerant plants can endure severe disease without suffering significant losses in quality or yield; however, these tolerant plants do not significantly inhibit the pathogen's activity, and disease symptoms may be clearly evident. Resistant plants usually suppress the pathogen in some fashion. A third term, immunity, means that a particular plant host simply cannot be infected by a specific pathogen. The garlic rust fungus (*Puccinia allii*) may infect several *Allium* species but has no ability to affect nonalliums such as peppers. Pepper crops, therefore, are immune to the fungus. Organic production systems are only permitted to use classically bred cultivars—those that were generated without using any genetic engineering techniques or methods.

There are distinct advantages to planting disease-resistant plant cultivars. Such selections are completely nondisruptive to the environment. In fact, their use enables growers to reduce and in some cases eliminate the application of chemicals they might otherwise use for pathogen control. The use of cultivars that are resistant to one disease is compatible with disease management steps taken to control other diseases. A final advantage is that for some host-pathogen systems, resistance is stable and long lasting so the cultivar can remain resistant for many years.

There are limitations to the use of resistance as a disease management tool (table 5.2). The greatest shortcoming is that acceptable resistance is not available for all diseases on all crops. For several of the most damaging plant diseases, such as tomato late blight (*Phytophthora infestans*) and white rot (*Sclerotium cepivorum*) of alliums, no resistant cultivars are currently available. In addition, commercial seed companies and plant breeders rarely invest in efforts to develop resistant cultivars for specialty or minor crops. Hence, there are specialty commodities—many of which are popular choices for organic producers—that will continue to lack resistance to their disease problems. Even if resistant varieties are being developed, the long development time and high market demand for resistant cultivars will result in expensive seeds, and that will have an effect on farmers' budgets.

Another shortcoming of some resistant cultivars is that they lack adequate horticultural characteristics with regard to appearance, quality, color, yield, and other important criteria. Celery that is resistant to *Fusarium oxysporum* f. sp. *apii* may not succumb to Fusarium yellows fungus, but the plant may also be unacceptably ribby, short, or low yielding. A cultivar that is resistant to one disease may also be quite susceptible to another important disease or insect pest. A lettuce cultivar that is resistant to *Lettuce mosaic virus*

Table 5.2. Examples of types of resistance

Degree or nature of resistance	Examples
Satisfactory resistance is not available.	virus diseases for peppers, late blight for tomato, white rot for alliums, lettuce drop for lettuce, many diseases for specialty crops
Satisfactory resistance is available but horticultural features may be less than optimal.	Fusarium yellows of celery
New strains of the pathogen will likely overcome currently available resistant cultivars.	downy mildew of lettuce, downy mildew of spinach
A cultivar resistant to one pathogen is very sensitive to other pathogens.	lettuce resistant to *Lettuce mosaic virus* is sensitive to the corky root pathogen

may be sensitive to corky root disease (caused by *Rhizomonas suberifaciens*); a lettuce selection that resists corky root may be very susceptible to downy mildew *(Bremia lactucae)*. One final disadvantage to resistance is that, depending on the host-pathogen system, resistance may not be long lasting: New strains of the pathogen may readily develop, rendering the crop susceptible once again.

Depending on which particular disease is involved, the failure of plant resistance can be either a rarity or a regular event. In most cases, resistance failure is attributed to the development of new strains of the target pathogen that overcome the host cultivar's resistance genes. Downy mildew disease of spinach provides a good case study of this phenomenon. From 1989 to 2008 in California, a new race of the spinach downy mildew fungus *(Peronospora farinosa* f. sp. *spinaciae)* would periodically occur, causing significant damage to the previously resistant spinach cultivars. Plant breeders countered with new cultivars containing genes resistant to the pathogen's new race. Growers then enjoyed several years of mildew-free spinach until eventually another race developed. This back-and-forth dynamic occurred for every one of the five downy mildew races that were confirmed in California during this period.

Despite the challenges of developing resistant cultivars and the setbacks of resistance breakdown, resistant plants remain among the most important weapons for disease control in organic systems.

Table 5.3. Keys to selecting sites

· Keep detailed records of which pathogens are confirmed at a particular site.

· Avoid locations with a history of soilborne fungi and bacteria: *Armillaria, Fusarium, Plasmodiophora, Sclerotium, Verticillium, Rhizomonas.*

· *Phytophthora, Pythium,* and *Rhizoctonia* often are much more widespread, so site selection might be less useful in avoiding these organisms.

· Nematodes (cyst nematode, root knot nematode) and some viruses *(Lettuce necrotic stunt virus)* can also be soilborne. Therefore, make a record of such problems if they develop.

· Note where significant outbreaks of virus and viruslike diseases occur (e.g., *Aster yellows, Tomato spotted wilt).* Avoid planting crops that are sensitive to these problems in such areas.

· Even foliar diseases (e.g., lettuce downy mildew) can be more severe in areas that have favorable microclimates. Make note of such occurrences and consider them when planning a planting schedule.

Organic growers are encouraged to actively and thoroughly investigate which resistant cultivars are available and to determine which cultivars perform best under their particular growing conditions.

SITE SELECTION

Before putting crops in place, a grower should carefully plan out planting and crop rotation strategies to try to avoid any known problem areas (see table 5.3). A grower can incur significant losses if he or she plants susceptible crops into a field known to be infested with persistent soilborne pathogens. Plant-pathogenic fungi such as *Armillaria, Fusarium, Plasmodiophora, Sclerotium,* and *Verticillium* are true soil inhabitants and will persist in soil for many years, even in the absence of a plant host. Not all fields are infested with these fungi, though, so growers are advised to select a planting site away from such fields. Soilborne pathogens such as *Phytophthora, Pythium,* and *Rhizoctonia* often are much more widespread, so site selection might be less of an option in avoiding these organisms.

There are also other planting situations that create risks that you should avoid. Pastures, foothills, riverbanks, grasslands, and other areas that support weeds and natural vegetation often are reservoirs of pathogens that cause virus and viruslike diseases. The vectors that carry such pathogens can also be found in these high-risk areas and often migrate into nearby production fields. For example, the *Aster yellows* phytoplasma and its leafhopper vector can be found in weedy grasslands in coastal California. Once the grassland vegetation dries up in the summer, the leafhoppers migrate into adjacent lettuce or celery fields, resulting in *Aster yellows disease* in these fields.

Consider pertinent environmental factors when selecting a planting site. Crops planted very close to the seacoast tend to be more at risk from downy mildew as a result of increased and persistent humidity. Just a few miles inland, however, humidity can be significantly lower, decreasing the disease pressure. Knowledge of soil features is critical when you want to avoid certain root and crown diseases. A site that has well-drained, sandy soil reduces the risk of damping-off and root rot for sensitive crops such as spinach.

Careful and detailed record keeping is essential for any site selection decision. Growers should keep notes on previous soilborne disease problems associated with certain fields, the position of the fields in relation to other key areas (weed reservoirs), important environmental characteristics for each location, and the nature of soil, water, and other physical features of each site.

EXCLUSION

The practice of keeping out any materials or objects that are contaminated or infected with pathogens and preventing them from entering the production system is known as exclusion (table 5.4). For some diseases, seedborne pathogens are a primary means of dissemination. Growers should purchase seeds that have been tested and certified to be below a certain threshold infestation level or that have been treated to reduce pathogen levels. Note that the designation "pathogen-free seed" really is not valid, since it is not possible to know whether a seed lot is, in its entirety, absolutely free of all pathogens. Seed tests only examine representative samples, but in most cases the tests are accurate enough to give a true picture of the risk of diseases initiated by seedborne pathogens. If a grower produces or purchases transplants, the plants should be as free as possible from pathogen contamination (symptoms may or may not be visible) and from disease (symptoms will be visible).

When you are producing transplants or greenhouse crops, all materials should be clean and free of pathogens. By using clean or new pots, trays, and soilless potting mix, a grower can prevent the introduction of soilborne pathogens to the greenhouse system. The recycling of potting mix is strongly discouraged, and pots and trays should be reused only after they have been properly cleaned with steam, bleach, or other disinfectants.

Soil and water can harbor pathogens as well. Take care to see that no infested soil or water is introduced into uninfested areas. *Lettuce necrotic stunt virus* of lettuce is found in river, flood, and runoff waters. Growers who have dredged up soil from ditches and dispersed it onto fields have found that their fields can become infested with the virus and that subsequent plantings can be severely diseased. Water draining from fields can carry a number of pathogens, so growers should not recycle or reuse it without carefully considering the potential risks and taking appropriate safety precautions. Soil and mud adhering to tractor equipment and implements can spread soilborne pathogens from infested fields into clean fields. Reduce the off-site movement of these infested materials as much as possible.

APPLYING CONTROL MATERIALS

Once vegetable crops are in the field or greenhouse, you may sometimes find it necessary to apply some sort of protectant or eradicant spray or dust for disease control, if any is available. Unfortunately, the selection of effective, proven materials that have been approved for organic use is limited (table 5.5).

Mineral-based disease control materials—primarily copper- and sulfur-containing fungicides and hydrated lime—have been used for centuries. These products are generally inexpensive and widely available, and they pose only minimal threats to the environment. However, their efficacy for disease control varies. While protectant copper fungicides have some activity against a wide range of fungal and bacterial pathogens, they do not reliably provide complete control, so growers should not depend on them. Sulfurs also exhibit some activity against many pathogens, but they usually only provide high levels of control against certain pathogens,

Table 5.4. Keys to exclusion

· Use seeds that have been tested or treated for seedborne pathogens.
· Carefully inspect transplants for symptoms of disease.
· When possible, avoid bringing contaminated soil, water, and equipment into uninfested areas.

Table 5.5. Keys to disease control materials

· Realize that many of these materials provide little or no control of pathogens.
· Thoroughly test these materials to see whether they are helpful in your program.
· Be sure to leave untreated control areas when evaluating such materials.
· Ensure that the materials are acceptable for organic certification and to your markets.
· Apply materials in a timely manner and with thorough coverage.

such as powdery mildews. Both coppers and sulfurs can burn sensitive vegetable crops under some environmental conditions, so use of these materials requires special care.

Horticultural oils, plant extracts (botanicals), minerals (such as kaolin), hydrogen peroxide, and other natural products are being investigated and used to control diseases. These products should be compatible with organic production practices, but their reliable effectiveness for disease control has yet to be demonstrated.

Bicarbonate-based fungicides have recently become available for control of plant diseases. Bicarbonates have demonstrated acceptable activity against powdery mildew and a few other diseases. It is not known, however, whether bicarbonates alone can provide season-long protection for an organically grown crop.

Disease control using microorganisms, or biologicals (biocontrol), or chemical by-products made by microorganisms is generating a good deal of interest. However, the history of successful biological control of plant diseases is not encouraging. Very few effective, economically feasible biological control materials are commercially available. Much research and development remains to be done.

Compost teas (fermented water extracts from composts and other organic materials) are showing some activity against select pathogens. Such spray materials are produced by soaking the organic substrates in water and then extracting the liquid by steeping and filtering. The resulting liquids contain a great variety of beneficial microorganisms, nutrients, and other chemicals. Foliar applications of compost teas have provided some field control of apple scab, powdery mildew, and gray mold. However, due to heightened concerns over food safety, compost teas must be carefully made and carefully used. These extracts have the potential to become contaminated with enteric bacteria, raising concerns about food product contamination.

For the best possible results with any of these materials, appropriate application technique (proper equipment, spray volume, and plant coverage) and timing are essential. Most materials do not perform well if the disease is already established, so applications should be made prior to the occurrence of any extensive infection. Before applying a product, a grower should confirm that the material is approved for use in organic production. Consult product labels, organic industry standards, UC Cooperative Extension farm advisors, pest control advisers, and your local Agricultural Commissioner's Office for product use information and restrictions.

CULTURAL PRACTICES

There are a number of cultural practices that a grower should consider when designing an integrated disease control system (table 5.6). As a general approach, growers should take steps to grow vigorous, high-quality plants using the best farming practices possible. Listed below are some specific cultural practices that can help to manage diseases.

Crop rotation. Crop rotation is an important tool for disease management. A rotation of diverse crops, inclusion of cover crops, and appropriate use of fallow (host-free) periods all can contribute to the reduction of inoculum levels for soilborne pathogens and the increase of diversity in soil microflora. In contrast, consecutive plantings of the same crop in the same field often lead to increases in soilborne pathogens. Too little crop rotation in a field can also simulate a monoculture effect that might increase foliar diseases.

Recent research has shown that certain plants that are revenue-generating crops also have a suppressive effect on diseases. For example, after broccoli is harvested and the plant residue is plowed into the soil, the decomposition of the

Table 5.6. Examples of cultural practices

- Rotate crops.
- Plant crops and cover crops that directly reduce soilborne pathogens.
- Avoid planting crops and cover crops that might increase soilborne pathogens.
- Time the planting of crops so that plant development is favored and pathogen growth is limited.
- Prepare proper seed and transplant beds to enhance plant germination and growth.
- Implement appropriate irrigation systems.
- In special cases, use lime and certain fertilizers to assist in disease suppression.
- Practice sanitation to remove sources of the pathogen.
- Control weeds.
- Manage insects.
- In special cases, use soil solarization and reflective mulches to assist in disease suppression.
- In greenhouses, manipulate the environment to disfavor pathogen development.

broccoli stems and leaves releases chemicals that can significantly reduce the number of *Verticillium dahliae* microsclerotia in the soil. This effect, which also holds true for other cruciferous crops, can be an important factor to consider when planning crop rotations. Some cover crops (e.g., mustards, sudangrass) may also share this beneficial effect and so could be considered in the crop rotation scheme. It is important to remember that while rotations with nonsusceptible plants and cover crops may help reduce soilborne pathogen numbers, any significant decrease in pathogen populations is likely to take many seasons.

When devising a crop rotation strategy, a grower should also be aware of which crops and cover crops might tend to increase disease problems. A vetch cover crop planted into a field with a history of lettuce drop can greatly increase the number of infective sclerotia of *Sclerotinia minor*. Vetch is a known host of root knot nematode (*Meloidogyne* species) and also might increase soil populations of *Pythium* and *Rhizoctonia* damping-off pathogens. While oilseed radish could be a potential trap crop for cyst nematode (*Heterodera* species), as a cover crop it is a host for root knot nematode and the clubroot organism (*Plasmodiophora brassicae*).

Time of planting. There are many factors to consider in regard to planting a crop. Timing can be an important question. If you plant cauliflower into a *Verticillium*-infested field in the spring or summer, it is likely to be subject to disease and, possibly, crop loss. However, if cauliflower is planted into the same field in the late fall or winter, it will exhibit no Verticillium wilt symptoms, presumably because the soil is too cool to allow the fungus to develop and cause significant disease.

Soil preparation. You can also reduce the incidence of disease by taking certain steps before and during planting. Preplant tillage should be thoroughly conducted so as to encourage the breakdown of plant residues from the previous crop. Proper preparation of the field and of the subsequent raised beds should reduce problems in areas that would otherwise be subject to poor drainage, pooling of water, and other conditions that favor pathogens. Soil and bed preparation should improve soil tilth to favor plant development for seed or transplants. Planting depth should be tailored to enhance seed emergence or transplant establishment. Poor soil preparation can result in stressed and exposed plants and increased damping-off problems from soilborne pathogens.

Irrigation. Irrigation management is clearly an important factor when it comes to disease control. Regardless of which irrigation system you use (furrow, sprinkler, or drip), the timing and duration of irrigations should satisfy crop water requirements without applying excess water. Overwatering greatly favors most soilborne pathogenic organisms. Most foliar diseases are worse with overhead sprinkler irrigation, which tends to enhance pathogen survival and dispersal as well as disease development. Bacterial foliar diseases are particularly dependent upon rain and sprinkler irrigation. Consider limiting or eliminating sprinkler irrigation if foliar diseases are problematic for a particular crop or field.

Fertilizers. The selection and application of fertilizers can, in a few documented situations, significantly influence disease development. For example, the use of the nitrate form of nitrogen fertilizers can increase the severity of lettuce corky root disease. The excessive use of nitrogen fertilizers can result in leaf growth that is overly succulent and more susceptible to some diseases. Other treatments can restrict disease activity. For instance, liming the soil to raise pH levels can reduce symptom expression for clubroot disease of crucifers. In general, however, fertilizer management is not directly related to disease control.

Field sanitation. In this discussion, field sanitation is the removal or destruction of diseased plant residues from a field. In some field situations it is a useful practice for managing diseases. Once lettuce has been harvested, for example, the remaining plant parts can serve as a reservoir for lettuce mosaic virus. Sanitation in this case would include plowing down the old plants. Lettuce drop, caused by the fungus *Sclerotinia minor*, occurs when sclerotia develop on lettuce plant residues and remain in the top few inches of soil. One form of sanitation involves deep plowing with moldboard plows, which invert the soil and bury the sclerotia. This procedure is only effective, though, for low to moderate sclerotia populations.

Sanitation measures are more common in greenhouse situations. The removal of dead or dying transplants can help reduce inoculum that could otherwise spread to adjacent transplants. The removal of senescent tomato or cucumber plants might reduce (though not prevent) the spread of *Botrytis* spores. Roguing is a special form of plant sanitation that involves the physical removal of diseased plants from the field. While there are many situations in which roguing will not work, researchers have shown that for sclerotia-forming fungi such as *Sclerotinia minor* on lettuce the regular removal of diseased plants can gradually reduce the overall number of sclerotia in the field.

The management of other pests is an aspect of cultural operations that can greatly influence the development of plant diseases. For example, virus disease management is more effective when weeds and insects are controlled. Weeds are known reservoirs of a number of viral and bacterial pathogens that can infect vegetable crops.

Soil solarization. In soil solarization, a grower covers the soil surface with plastic tarps to increase soil temperatures to levels that kill soilborne pathogens, weeds, and other crop pests. Soil solarization works best in areas with high summer temperatures, conditions that generally do not occur in California's coastal regions. Soil solarization will not eradicate a pathogen from a field, but it may lower pathogen populations. Soil flooding is a related though seldom-used means of creating conditions—in this case, saturated soil over an extended period—that may result in a decline of soilborne pathogens.

Reflective mulches. Another tool for controlling the incidence and severity of virus diseases is reflective mulch, a polyethylene sheet with an aluminum top layer. These specialized tarps can repel insects such as aphids and so reduce the number of virus vectors that land and feed on the plants. In addition, researchers have found that, while reflective mulches do not eliminate virus diseases, in some cases the mulches delay disease development and the expression of symptoms.

Finally, a grower or nursery operator can manipulate environmental conditions within a greenhouse vegetable transplant or production system to help control diseases. Botrytis diseases can be better managed if warm, humid air is vented out of the greenhouse. Because rain is not a factor in greenhouses, many bacterial foliar diseases can be virtually eliminated if the grower uses a drip or sub-irrigation system.

COMPOSTS

The incorporation of composts into the soil is a fundamental cultural practice in organic vegetable production. Composts benefit the soil's fertility and condition in a number of ways and probably also help with disease management. However, there is a lack of research studies and empirical data that clearly document any direct disease control benefits from compost application. Even so, it is a good idea to add composts to farmed soils in order to increase soil microflora diversity and populations.

PLANT DISEASE DIAGNOSTICS

The first step in any management decision regarding disease control is to determine which diseases and pathogens are causing the problem (table 5.7.). Accurate and timely diagnosis of plant diseases is an essential component of integrated disease control in organic and conventional systems. Disease diagnosis is enhanced when all professionals—including the grower, field personnel, pest control adviser, consultant, and Extension personnel—work together to ascertain the cause of the problem. Often, disease identification in the field is impossible, and you have to submit samples to a qualified laboratory for analysis.

Table 5.7. Key steps and information needed for plant disease diagnosis

Category	Key steps and strategies
Key strategies for effective plant disease diagnosis	Develop and use a systematic, organized method.
	Keep detailed, complete records.
	Make special note of the development of soilborne diseases.
	Regularly assess crop status throughout the production cycle.
	Collect symptomatic plants as they occur; have the samples analyzed by a lab.
	Develop host range lists for the various pathogens.
	Work with your team of UC Cooperative Extension Farm Advisors, pest control advisers (PCAs), and other professionals.
Gathering information about the plant host	What are the characteristics and growth habits of a healthy plant?
	What is the cultivar of the plant in question? Is the cultivar suited to the production area? Is the cultivar particularly susceptible or resistant to diseases or other problems?
	What symptoms are present on the affected plant parts? Be sure to examine all parts of the plant, especially the root.
	What is the distribution of the symptoms on any one particular plant?
	Do symptoms occur only on one side of the plant? Only on older leaves? Only on younger leaves? On taproots but not on feeder roots?
	What is the distribution of diseased plants in the field? Are affected plants found randomly in the field or are discernible patterns present?
	What is the growth stage of the affected plant? Seedling? Mature? Senescent?
	If the plant being examined is a sample brought in from the field, what is the condition of the sample? (Samples in poor condition may not yield useful information.) Is the sample representative of the field problem? Have unaffected plants also been included in the sample for comparison?
Gathering information about the possible pathogen	As background information, which diseases are known to occur on the host?
	Are these diseases known to occur in the geographic area of concern?
	Are there signs of the pathogen (is it visibly present)?
	What is the distribution of signs on the affected plant? On all plant parts or only on certain parts?
	Are there multiple signs that indicate more than one pathogen?
	If signs are present, can the organism be identified in the field (for example, mildews, rust, cyst nematode)?
	Can the pathogen be identified after incubation, isolation procedures, microscopic examination, serological tests, or other laboratory tests?
	When the agent is identified, what is the host range of the pathogen and what are possible sources of inoculum?
Gathering information about the surrounding environment	What time of year did the problem occur? Early or late in the season?
	What are the current and past weather conditions?
	What is the general location of the field (coastal, inland, etc.)?
	What is the location of the field in relation to roads, other crops, buildings, water sources, weeds, etc.?
	Are there other plants (nearby crops, weeds) in the area that show similar symptoms or that could provide a source of pathogen inoculum?
	What are the soil type and condition (heavy clay, porous sand, compacted layers, change in soil type from affected to unaffected areas, etc.)?
	What is the water source and what is the quality of the water?
	How does the water drain off the field? Are there low spots where water collects?
	What environmental conditions favor or limit host plant development?
	What environmental conditions favor or limit pathogen and disease development?
	Is there evidence of other biotic stress factors (insect/invertebrate damage, competition from weeds, vertebrate damage, etc.)?
	Is there evidence of abiotic stress factors (temperature extremes, moisture stress, salt buildup, mineral deficiencies or toxicities, pollution, wind or other mechanical damage, etc.)?
Gathering information about production practices	What are the normal, typical production practices for the crop (planting or transplanting, irrigating, fertilizing, cultivating, pruning, pest management)?
	What procedures have been completed or omitted for the crop in question?
	What fertilizers, pesticides, etc., have been applied to the crop?
	Is there any indication of misses or skips in the application of chemicals?
	What is the previous crop history of the field in question?
	For any previous crop, what were the production practices that were completed or omitted (including use of herbicides and other pesticides)?
	What production practices have taken place in adjacent areas? Is there any indication that adjacent applications of chemicals have resulted in the drift of applied material onto the field in question?

Once you have a diagnosis, you can settle on appropriate steps to manage the problem. Again, detailed record keeping will help you deal with the current problem and at the same time build a database that you can use to plan disease management steps for future crops.

For assistance with plant disease diagnosis or help in finding a testing laboratory, contact your pest control adviser, local UC Cooperative Extension Farm Advisor, or other professionals trained in plant pathology, pest management, or plant production.

REFERENCES/RESOURCES

Dillard, H. R., and R. G. Grogan. 1985. Influence of green manure crops and lettuce on sclerotial populations of *Sclerotinia minor*. Plant Disease 69:579–582.

Gaskell, M., et al. 2000. Organic vegetable production in California: Science and practice. HortTechnology 10:699–713.

Gliessman, S. R. 1995. Sustainable agriculture: An agroecological perspective. In J. H. Andrews and I. Tommerup, eds. Advances in Plant Pathology, 11:45–57. San Diego: Academic Press.

Irish, B. M., J. C. Correll, S. T. Koike, and T. E. Morelock. 2007. Three new races of the spinach downy mildew pathogen identified by a modified set of spinach differentials. Plant Disease 91:1392–1396.

Koike, S. T., et al. 1996. Phacelia, lana woollypod vetch, and Austrian winter pea: Three new cover crop hosts of *Sclerotinia minor* in California. Plant Disease 80:1409–1412.

Koike, S. T., R. F. Smith, and K. F. Schulbach. 1992. Resistant cultivars, fungicides combat downy mildew of spinach. California Agriculture 46:29–31.

Leggett, M. E., and S. C. Gleddie. 1995. Developing biofertilizer and biocontrol agents that meet farmers' expectations. In J. H. Andrews and I. Tommerup, eds. Advances in Plant Pathology, 11:59–74. San Diego: Academic Press.

Mathre, D. E., R. J. Cook, and N. W. Callan. 1999. From discovery to use: Traversing the world of commercializing biocontrol agents for plant disease control. Plant Disease 83:972–983.

McGee, D. C. 1981. Seed pathology: Its place in modern seed production. Plant Disease 65:638–642.

———. 1995. Epidemiological approach to disease management through seed technology. Annual Review of Phytopathology 33:445–466.

Pankhurst, C. E., and J. M. Lynch. 1995. The role of soil microbiology in sustainable intensive agriculture. In J. H. Andrews and I. Tommerup, eds. Advances in Plant Pathology, 11:229–247. San Diego: Academic Press.

Scheuerell, S., and W. Mahaffee. 2002. Compost tea: Principles and prospects for plant disease control. Compost Science & Utilization 10:313–338.

Severin, H. H. P., and N. W. Frazier. 1945. California aster yellows on vegetable and seed crops. Hilgardia 16:573–596.

Subbarao, K. V., J. C. Hubbard, and S. T. Koike. 1999. Evaluation of broccoli residue incorporation into field soil for Verticillium wilt control in cauliflower. Plant Disease 83:124–129.

Subbarao, K. V., S. T. Koike, and J. C. Hubbard. 1996. Effects of deep plowing on the distribution and density of *Sclerotinia minor* sclerotia and lettuce drop incidence. Plant Disease 80:28–33.

van Bruggen, A. H. C. 1995. Plant disease severity in high-input compared to reduced-input and organic farming systems. Plant Disease 79:976–984.

Weed Management for Organic Vegetable Production in California

RICHARD F. SMITH, W. THOMAS LANINI, MILTON E. MCGIFFEN, JR., MARK GASKELL, JEFF MITCHELL, STEVEN T. KOIKE, AND CALVIN FOUCHE

To achieve economically acceptable weed control and crop yields, an organic vegetable grower must use a variety of techniques and strategies. Weeds can always be pulled or cut out, but the question is, simply, how much time and money is it reasonable for a grower to spend on reducing weed pressure? The more a grower is able to reduce weed seeds and perennial propagules in the soil, the more economical crop production will be.

A grower's ideal solution would be to keep the farm weed-free. In practice this goal may not be possible, but any reduction in weeds and in the number of weed seeds or perennial propagules that reach the soil will make subsequent weed control operations less expensive. Organic weed management requires a two-pronged strategy. There is a long-term component that includes broad weed impacts, such as the type of weeds present, types of crops and crop rotations, timing of tillage operations, consistency of effective weed management, availability of effective weed control resources, and the overall impact of those measures on the soil's seed bank. If long-term strategies are successful over time, they can reduce weed pressure and subsequent weeding costs. Short-term weed control strategies deal with techniques and practices employed to reduce weeding costs on the current crop.

A good understanding of how weed infestations become established and what resources weeds require is useful when formulating a management plan. Like other plants, weeds require water, nutrients, and light. The first or biggest plant to occupy a site has a competitive advantage over other plants at that site. The cultural practices used to produce vegetables (i.e., the use of transplants, pre-emergent flaming of weeds, pregermination of weeds, etc.) can allow the crop to quickly establish and so gain a competitive advantage over weeds. The goal is to help the crop outcompete weeds and reduce the availability of resources to them. If organically acceptable cultural techniques

can give the crop a competitive advantage, subsequent hand-weeding operations and costs are less. In this chapter we arrange organically acceptable weed control strategies into a few broad categories and discuss many of the common short- and long-term techniques available to organic growers to manage weeds in a vegetable production operation.

CULTURAL PRACTICES

Crop rotation. The type of vegetable crop you grow and the other crops you rotate into the field both affect weed pressure. For instance, the culture of short-term, intensively cultivated crops, such as lettuce, helps to manage perennial weed species such as bindweed (*Convolvulus arvensis*) and annual weeds with a long life cycle, such as little mallow (*Malva parviflora*). On the other hand, short-term crops also tend to favor weeds that can set seed rapidly between planting and cultivation operations, such as groundsel (*Senecio vulgaris*) and burning nettle (*Urtica urens*). Longer-season vegetables, such as peppers and tomatoes, make control of perennial

ECONOMIC VALUE OF WEED CONTROL

Celery is a high-value vegetable crop, and weed control is a big part of its production costs (figure 6.1). In one study the grower would have lost more than $3,237 per acre if he had not controlled weeds. While drip irrigation was more expensive than sprinklers, the drip system wetted less of the ground and reduced weed seed germination. The reduced weed control costs increased overall profits when the grower used drip irrigation.

Returns are generally lower when hand-weeding alone is used as compared to when hand-weeding is combined with herbicide use in conventional production. Organic growers can overcome the higher costs related to additional hand-hoeing by setting higher market prices for certified organic produce.

weeds such as field bindweed and yellow nutsedge (*Cyperus esculentus*) more difficult because these weeds resprout after early-season weed control operations and so replenish their root and nutlet reserves (see sidebar, "Economic Value of Weed Control"). Crop rotations can serve as a long-term control practice for problematic weeds, but for most growers weed control is only one aspect to consider when making crop rotation decisions. Still, it remains an important weed management tool.

Water management. Water management is a key tool for controlling weeds in a vegetable operation. Careful irrigation management can reduce weed populations in a number of ways:

1. *Pregermination of weeds.* Pregermination involves the use of irrigation or rain to stimulate weed seed germination before you plant the cash crop. The weed seedlings are allowed to emerge but are then killed by shallow cultivation, flaming, an organic herbicide application, or a combination of these treatments, before the crop is planted. Pregermination should take place as close as possible to the crop planting date to ensure that the weed spectrum will not change before the vegetable crop is planted. Changes in the weed spectrum may be caused by changes in the season or weather. The time of year, type of irrigation system, and interval between irrigation and weed control all affect the efficacy of this technique. A wait of 14 days between pre-irrigation and shallow tillage for weeds to emerge provided from 33 to 65 percent control, and sometimes more (table 6.1 and figure 6.2) in the Central Coast lettuce crop studied by Shem-

Tov, Fennimore, and Lanini (2006). If time permits, you can repeat the pre-germination cycle to further reduce weed populations. The effect can be augmented by flaming or organic herbicide treatments used to kill the flush of weeds anytime between crop seeding and emergence. Timing of treatment is important for all weed control methods: Small weed seedlings (fewer than two true leaves) are easier to kill than large weeds. Note that flaming and organic herbicides are less effective on grass weeds.

2. *Planting to moisture.* A similar technique is planting to moisture. For this technique, you apply water to the field, allow weeds to germinate, cultivate to kill the germinating weeds, and then allow the top

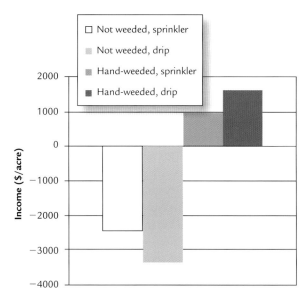

Figure 6.1. Where weeds were controlled in a celery crop, drip irrigation wetted less of the soil surface than sprinkler irrigation, decreasing weed seed germination and, because of reduced weed control expenses, netting a higher profit for the grower. *Source:* Ogbuchiekwe and McGiffen 2001.

Table 6.1. Effect of pre-irrigation on weeds in the subsequent lettuce crop and hoeing time for weeds in lettuce (2003)

Preplant irrigation treatment	Spring lettuce		Fall lettuce	
	Weed density (weeds/sq ft)	Weeding time (hr/acre)	Weed density (weeds/sq ft)	Weeding time (hr/acre)
No pre-irrigation	30 a*	13.9 a	15 a	12.8 a
Furrow pre-irrigation	20 b	12.3 b	7 b	10.5 b
Sprinkler pre-irrigation	16 b	10.5 c	5 b	0.2 b

Note: Mean separation within columns is according to Fisher's LSD test (P=0.05) and is indicated by letters, with similar letters indicating nonsignificant difference.

Source: Shem-Tov, Fennimore, and Lanini 2006.

Figure 6.2. Weeds in lettuce (left) with no preirrigation and (right) with preirrigation followed by shallow tillage two weeks later. *Photos:* Steve Fennimore.

Figure 6.3. Moisture (left) below the dry dust mulch and beans (right) emerging through the dust mulch. *Photos:* Richard F. Smith.

2 to 3 inches of soil to dry and form a dust mulch. Large-seeded vegetables, such as sweet corn, beans, or squash, can be planted deep (1.5 to 3 inches) into the moist, underlying soil. These crop seeds then germinate, grow up through the dust mulch, and become established before there is any need for supplemental irrigation, thus getting a head start on weeds (figure 6.3).

3. *Buried drip irrigation.* If buried below the surface of the bed, drip tape provides moisture to the crop roots (figure 6.4) and minimizes the amount of moisture available to germinate weeds on the soil surface (figure 6.5). Drip irrigated crops generally have lower weed populations and reduced weed control cost, especially during the nonrainy periods of the year (see figure 6.1).

Crop competition. Vigorously growing crops can often out-compete weeds. Weeds grow best in areas where competition is sparse, such as between rows or in gaps in the crop stand. You can increase the crop density by reducing the in-row spacing between plants or placing rows closer together and so minimizing the space available for weeds to grow. This improves the crop's competitiveness by allowing the crop to close its canopy sooner. Some crops (e.g., tomatoes, beans, and sweet corn) establish competitive canopies and compete effectively with

Figure 6.4. Buried drip irrigation tape provides moisture to the roots of the crop but keeps the soil surface dry and reduces weed pressure. *Photo:* Richard F. Smith.

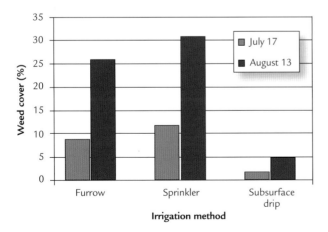

Figure 6.5. Impact of the type of irrigation system on weed growth. *Source:* Grattan et al. 1988.

weeds if given an early competitive advantage, whereas others (e.g., onions and garlic) do not compete well with weeds. The use of transplants planted into a weed-free bed, though, gives any vegetable crop an early competitive advantage over weeds. The elimination of weeds with cultivation and hand-weeding just before the crop canopy closes can provide excellent weed control in many crops for the remainder of the growth cycle.

Reducing the weed seed bank. The old saying "One year's seeding means seven years' weeding" reminds us of the consequence of letting weeds go to seed. Practices that reduce the contribution of weed seeds or vegetative propagules to the soil seed bank reduce weed pressure over time and can reduce weeding costs in the long run. Ideally, no weeds should ever be allowed to go to seed, because that can aggravate weed problems for many years in the future. As an example, the seed of common purslane remains viable in the soil for 40 years (Darlington and Steinbauer 1961). The longevity of weed seeds, when considered together with the large numbers of seed produced by an individual plant (e.g., 100,000 seed per plant for purslane or barnyardgrass), can lead to long-term buildup of enormous seed banks in the soil. Diligent weed control during the growing season as well as the fallow period reduces weed seed production and can reduce weed pressure over time.

As mentioned above, certain crops such as short-season lettuces can provide opportunities for frequent cultivation and quick rotations that reduce the ability of some weeds to mature and set seed. Highly competitive cover crops can also smother weeds before they set seed. If you carry weeds that have mature seed out of the field during hand-weeding operations and dispose of them away from the field, that can also significantly reduce the seed bank (figure 6.6). Another potential source of weed seeds is manure compost, and care must be taken to insure that such materials are well composted before they are applied to make sure that they do not introduce additional weed seeds into the field.

Cover crops. Cover cropping is a key cultural practice in organic production that provides a variety of benefits for the crop. However, cover crops do have the potential to increase weed pressure in vegetable production systems by allowing weed seed production during the cover crop production cycle. Weed plants often decompose before the end of the cover crop cycle so that you might not ever notice that weeds had been in the cover crop. In such a case, the cover crop acts as a nurse crop to weeds, allowing them to mature and make a substantial contribution to the seed bank. For instance, many legumes and cereal-legume cover crop mixes planted at a normal seeding rate (i.e., 100 to 120 lb/acre) are not competitive and allow substantial weed growth and seed set early in the cover crop's growth cycle (see table 6.2 and sidebar, "Nutsedge Management"). Cover crops that compete effectively with weeds have rapid growth early in their season and cover the soil surface within 30 days. An adequate seeding rate is also an important factor in providing rapid ground cover and smothering out weed seedlings. Competitive cover crop varieties include Merced rye, white mustard (*Sinapis alba*), Indian mustard (*Brassica juncea*), and higher seeding rates of legume-cereal mixtures (i.e., >200 lb/acre).

A grower can use implements such as a flex-tined or rotary hoe cultivator to cultivate cereal and cereal-legume mix cover crops when weeds are at the white stem stage (15 to 30 days after seeding the cover crop). Rotary hoeing reduced weed seed production by chickweed and shepherd's purse by 80 to 95 percent (figure 6.7) in a legume-cereal cover crop in a 2-year study conducted in Salinas (Boyd and Brennan 2006). This technique takes advantage of the difference in the depth of the seedling emergence between the weeds (shallow) and the cover crop (deeper) and can reduce weed pressure in cover crops without substantially reducing cover crop yield.

NUTSEDGE MANAGEMENT

Purple and yellow nutsedge are two of the world's worst weeds. The nutsedges are problem weeds throughout California, with purple nutsedge predominating in the southern half of the state. These perennials have a remarkable ability to survive adverse conditions and then grow explosively when the land is planted to irrigated crops (figure 6.8). Losses can result when nutsedges compete with crops, thereby decreasing yield, or when they directly damage belowground plant parts such as onion bulbs. Nutsedge can even decrease property values, because potential renters or buyers of the land know that, once established, it is nearly impossible to eradicate the weed.

The nutsedges produce both seed and tubers, but most reproduction is by tubers. The tubers may be thought of as a resting stage that allows the weeds to survive adverse conditions. Many people say tubers can survive almost anything. In fact, a large percentage of tubers often die during dormancy, but as little as 1 percent of the tubers from a previous infestation is more than enough to bring back the population of these prolific weeds.

Understanding nutsedge control begins with the realization that tubers are the key to the weed's survival. Prevent tuber production and you will eliminate the weed. Bear in mind, though, that the tubers can remain dormant and impervious to pesticides for 2 or possibly 3 years. Control programs should be aimed at preventing the formation of tubers by preventing the growth of nutsedge plants. If no new tubers are formed, tuber mortality will eventually eliminate nutsedge problems.

The best control of nutsedge is often obtained by growing competitive crops. Nutsedge is susceptible to shading, so crops that quickly form a dense canopy can outcompete the weed. Sudangrass, corn, and wheat are crops that may be able to shade out nutsedge. Conversely, melons and other warm-season crops that do not shade the ground may be overrun by nutsedge infestations. Sudangrass is a rotational crop that can be grown during the summer to shade out nutsedge and then followed with winter vegetables. The field in figure 6.9 was so heavily infested it looked as if a purple nutsedge lawn had been planted with sudangrass. By midsummer, the dense sudangrass canopy had killed emerged nutsedge.

Table 6.2. The effect of early-season ground cover by cover crops on the establishment of and eventual seed production by burning nettle

Cover crop	Percentage ground cover 28 days after planting	Burning nettle seed production (viable seeds per m^2)	
		Year 1	Year 2
Cayuse oats	30 a*	6,010 a	1,861 a
Legume-oats mix	18 a	13,622 a	4,282 a
Merced rye	46 b	—	275 b
Mustard blend	78 c	1,283 b	14 c

* *Note:* Mean separation within columns is according to Fisher's LSD test (P=0.05) and is indicated by letters, with similar letters indicating nonsignificant difference.
Source: Brennan and Smith 2005.

Figure 6.6. Bags of purslane plants removed from the field and carried to the edge of the field for disposal can help reduce the number of seed that are deposited to the seed bank. *Photos:* Richard F. Smith.

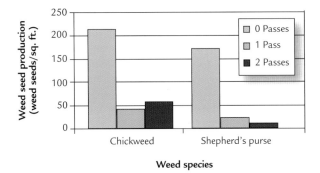

Figure 6.7. Weed seed production by chickweed and shepherd's purse 134 days after planting date and following 0, 1, or 2 passes with the rotary hoe when weeds were in the white stem stage. Zero-pass data show significantly more weed seed production than 1- or 2-pass data. *Source:* Boyd and Brennan 2006.

Figure 6.8. Nutsedge grows well in high temperatures, and can overrun broccoli, lettuce, carrots (shown), and other cool-season crops when they are planted in the late summer. *Photo:* Milton E. McGiffen.

CULTIVATION

Cultivation is probably the most widely used weed control method in organic vegetable systems. Mechanical cultivation uproots or buries weeds. Weed burial works best on small weeds, while larger weeds are better controlled by destroying the root-shoot connection or by slicing, cutting, or turning the soil to eliminate the root system's contact with the soil. Cultivation is effective against annual weeds, but it can also spread viable root pieces of field bindweed and tubers of yellow nutsedge. Cultivation typically involves cutting implements that are pulled parallel to the bed, leaving an uncultivated band around the seed line (figure 6.10). Effective cultivation requires good land preparation in order to ensure the precision and accuracy necessary to remove weeds and leave the crop plants standing. Shallow cultivation usually is best: It brings fewer weed seeds to the surface. Level beds allow greater precision in setting the depth of tillage. Cultivation requires relatively dry soil conditions, and subsequent irrigations should be delayed long enough to allow exposed weed roots to dry and prevent their rerooting into the soil. The

proper timing interval between cultivations depends on the speed of weed growth: In spring, 2 to 3 weeks between cultivations is often adequate, but in the fall or winter longer periods between cultivations may suffice. Cultivation is typically practiced while the plants are small and is curtailed when the canopy begins to close and equipment can no longer enter the field easily without damaging the crop. Later-season cultivations can also disturb crop roots or knock flowers and fruits from the crop.

The goal of cultivation is to cut weed seedlings as close to the seed row as possible without disturbing the crop. In most cases, precision cultivation can control weeds on 80 percent or more of the bed (figure 6.11). The weeds that remain in the uncultivated seed lines can be removed by hand or other mechanical means. The following are common cultivation implements:

- Knives of various shapes, beet hoes, and sweeps (figure 6.12) can be used to cut and uproot weeds on bed tops within 1 to 3 inches of the crop row. Knives and sweeps can be combined with reversed-disc hillers, which cut vining weeds such as field bindweed and move soil away from the crop row.

Figure 6.9. Sudangrass quickly forms a tall, dense canopy when temperatures are warm. The brown nutsedge leaves emerged at the same time as the sudangrass, but were overtopped and died from the lack of light at the bottom of the sudangrass canopy. *Photo:* Milton E. McGiffen.

Figure 6.10. Broccoli plants growing in the 4-inch-wide uncultivated band left after cultivation. *Photo:* Richard F. Smith.

Figure 6.11. Difference between beds before and after an effective and precise cultivation. *Photo:* Richard F. Smith.

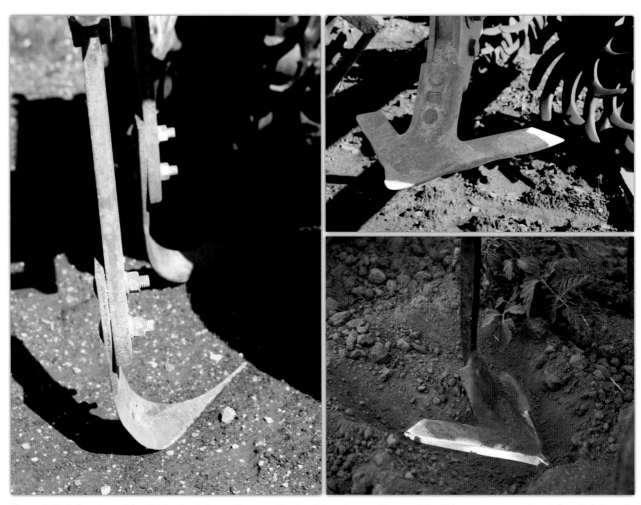

Figure 6.12. Left and top right: Cultivation knives and sweeps closely cultivate lettuce. Bottom right: This sweep cuts weeds as it is pulled through the soil. *Photos:* Richard F. Smith.

Figure 6.13. Rolling cultivators uproot weeds as the tines roll while the implement is pulled through the field. *Photo:* Richard F. Smith.

Shovels can be used to throw soil to the base of the plant, burying small weed seedlings in crops such as sweet corn and cauliflower that are grown in a single seed line per bed. Rolling cultivators (Lillistons) (figure 6.13), which are common cultivating implements for numerous crops, operate like rototillers but are ground driven. Rolling cultivators primarily uproot weeds, but they can also throw soil and bury weeds in the crop row.

- Some cultivators such as the spring-tine cultivators, torsion cultivators, and finger weeders are capable of removing weeds in the seed row (Bowman 1997). With care, these cultivators can take weeds out of the seed row on transplanted crops or crops with tough stems (e.g., sweet corn and green beans) but are not suited for direct-seeded crops such as lettuce and cole crops with delicate stems (Bowman 1997; Stone 2005a).

- Cultivation implements are often mounted on sleds to improve the accuracy and closeness of cultivation in row crops. Guide wheels, cone wheels, and other devices are also used, but cultivation sleds are generally the most precise. Knives and other cultivation implements are attached to the sled, which is pulled behind a tractor. The sled rides on the bottom of the furrow and facilitates greater precision and speed of cultivation. A tractor mounted with a belly cultivation bar can be used to conduct precise cultivation: This arrangement allows the driver to look down on and guide the cultivation operation. Computer vision guidance systems can increase the accuracy and speed of vegetable cultivation (figure 6.14). Growers can use them to guide pull-behind cultivation sleds and automatically adjust to the irregularities in the seed lines. With a computer vision guidance system, you can reduce the width of the uncultivated band to as little as 2 inches and take out a greater percentage of weeds from the bed (table 6.3).

Figure 6.14. Computer guidance system for a cultivator. The camera in the yellow box follows a seed line and signals to a computer when it needs to move the cultivation sled laterally to adjust for irregularities in the field. *Photo:* Richard F. Smith.

Table 6.3. Impact on weed control, weeding hours per acre, and yield of lettuce after cultivation that left a 2- or 4-inch-wide uncultivated band around the seed line

Cultivation band width	Weed control (%)	Weeding time (hr/acre)	Stand count per 100 ft of seed line	Bulk harvest (lb/acre)*
2 inches	56.8 a[†]	22.2 a	118.2 a	37,256
4 inches	25.0 b	30.3 b	117.2 a	38,569

Note: Commercial harvest data; not subjected to statistical analysis.

[†] Mean separation within columns is according to Fisher's LSD test (P=0.05) and is indicated by letters, with similar letters indicating nonsignificant difference.

Source: Smith and Bensen, 2005.

- Brush hoes use stiff brushes that sweep the soil surface and uproot weeds (figure 6.15). They can be set to cultivate narrow bands (i.e., < 2 inches), and they do not sheer or break the soil like traditional knives, so they may break fewer crop roots. Power-driven rotary cultivators with rototiller-type tines can remove larger weeds from between seed lines.

Perennial weeds such as field bindweed can be particularly problematic in an organic operation. If a serious infestation of perennial weeds prevents economic crop production, consider keeping the field fallow for a period and practice repeated cultivations at 2- to 3-week intervals to deplete the weed root system, reducing the weeds' vigor and survival. For field bindweed, set the cultivation knives to cut the plant 4 inches or more below the soil surface to maximize the depletion of the weed's root reserves.

Nighttime discing may help to reduce weed germination. This is because many weed seeds require a flash of red light to overcome dormancy and germinate. This flash occurs if buried weed seeds are tilled during the day, when the seeds are suspended with soil during tillage. In an Oregon

Figure 6.15. Stiff bristles of the brush hoe rotate and sweep away small weeds. The bristles allow a less aggressive treatment than traditional knives, which may shear the soil and break crop roots when cultivating close to the crop. *Photos:* Richard F. Smith.

study, weed germination increased 4- to 5-fold when the seeds were exposed to light in daylight tillage, as opposed to weeds tilled at night (Scopel, Ballare, and Radosevich 1994). After nighttime tilling, only those seeds left on the soil surface will germinate. That can still be quite a few seeds. Because these seeds are left on the soil surface, it may take several tillage operations before you see much effect. The Oregon study showed that most summer annual weeds like pigweed, lambsquarters, and barnyardgrass respond favorably to night tillage, as do many winter annual species. This effect may be offset somewhat in coarse-textured soils where daylight is able to penetrate deeper into the soil.

Deep plowing is a tillage technique that buries weed seeds or propagules of perennial plants so deep that they are unable to germinate. This technique is particularly effective on nutsedge tubers, whose viability rapidly declines within a year or two. However, the seed of most other weeds lasts many

years, and you can expect that they will be brought back to the surface by the next deep tillage. In addition, seeds buried deep in the soil are likely to last many times longer than shallow-buried seeds since animals are less likely to find and eat them.

FLAMERS

Weed flaming uses propane burners directed toward weeds on the beds (figure 6.16). Heat from the burner causes the plant's cell sap to expand, rupturing the cell membranes. This will occur in most plant tissues at about 130°F. You can use flaming after planting your crop but before crop emergence in slow-germinating vegetables such as peppers, carrots, garlic, and parsley. You can also use flaming as a directed weed control treatment at the base of sweet corn.

Flaming is most effective on broadleaf weeds with two or three true leaves or fewer. If you shield the

Figure 6.16. Propane flaming of weed seedlings causes the cell sap of the plants to expand and rupture cell membranes. *Photo:* Richard F. Smith.

flamers from wind, the heat will stay on the weeds longer for increased weed control. Grasses generally are not controlled by flaming because their growing point is below the ground. After flaming, the surface tissue of weeds that have been killed quickly changes from a glossy to a matte appearance. If the weed is properly flamed, your fingerprint should remain on a leaf after you press it between your index finger and thumb. Typically, flaming can be done while travelling at 3 to 5 mph through fields, although this depends on the heat output of the unit being used. Best results are obtained under windless conditions, as wind reduces the efficiency of the flamer. Evening and early morning are the best times to adjust the flame, since the flame is easier to see then than during daylight. Flaming does not reduce subsequent weed germination.

Propane-fueled flamers are the most common type. Infrared flamers use a burner to heat a ceramic surface and then direct the infrared radiation toward the target weed. While generally less effective than propane, infrared flamers do allow more precise placement of heat (Stone 2005a).

STEAM AND SOLARIZATION

Heat is used in organic agriculture to kill weeds. The heat is applied to the soil as steam or by soil solarization. For steam treatment, heated water generates steam that is injected into the soil, killing weed seeds. Large quantities of fuel and water are required for this kind of steam sterilization, and the expense restricts its use to small acreages of high-value crops.

Soil solarization traps heat from the sun beneath a layer of clear plastic, increasing the temperature in the top layer of soil to lethal levels. The water applied to the beds before the clear plastic is laid causes weed seeds to imbibe water; with the plastic in place, the trapped heat kills imbibed weed seeds or germinating weeds (figure 6.17). Solarization controls many species of weeds, but the depth to which it is effective depends on the temperature and duration of solarization. Hard-seeded weeds such as burr clover (*Medicago polymorpha*) generally are not controlled by solarization (Elmore et al. 1997). Soil solarization can significantly reduce the number of viable weed seeds in the top layer of the soil, and if soils are not disturbed following solarization the weed control can remain

Figure 6.17. The brown nutsedge inside this solarization tarp emerged before steam was produced but then died as the temperature rose. The green leaves on the outside are from the dense nutsedge infestation that would have been present if the bed had not been sterilized. *Photo:* Milton E. McGiffen.

effective for much of the season. This technique is used successfully in organic carrot production in California's desert and Central Valley regions, but its usefulness is limited on the coast due to foggy weather, which limits the availability of direct sunlight.

BIOFUMIGATION

Mustard cover crops (e.g., white and Indian mustards) release short-lived toxic chemicals called glucosinolates into the soil, and these can reduce weed emergence. In a 2-year study conducted on California's Central Coast, weed pressure was reduced by nearly half on a weedy site following incorporation of mustard cover crops (Smith 2004a). However, on low weed pressure sites it has been more difficult to observe and measure any significant reduction in weed pressure in response to biofumigation by mustards. The minimal impact of biofumigation on weeds may be due in part to the small amount of glucosinolates that mustards contain.

MULCHES

Mulch can prevent light from reaching the soil surface, thereby preventing weed germination and growth. Many materials can be used as mulches, including dark-colored plastic or organic materials such as cover crop residue, yard waste, straw, hay, sawdust, and paper products. Mulch needs to block all light to effectively control weeds. Different mulch materials vary in the thickness of application necessary to block light.

Plastic mulches laid over beds can provide weed control to a significant portion of the bed (figure 6.18). The most common colors of plastic used for weed control are black, brown, and green. All of these are effective at blocking the wavelengths of light needed for weeds to grow. Plastic mulches are laid over beds with their edges placed in the furrows, where they are covered with dirt to hold the mulch in place. Before laying the mulch, growers install drip irrigation to provide the crop with moisture. Certain weeds, such as nutsedge, have sharp leaves that can poke through the plastic, so they are not effectively controlled by colored plastic mulches. Weeds that emerge through the plastic at openings provided for crop growth must be removed by hand.

Figure 6.18. The black plastic mulch applied to the beds on the right can control weeds on a significant portion of the bed. However, weeds still need to be controlled in the holes cut for crop growth and in the furrows. *Photo:* Richard F. Smith.

Organic mulches, such as municipal yard waste, paper, cardboard, straw, hay, and wood chips, effectively control the germination of annual weeds if they are applied and maintained in a layer 4 or more inches thick. Organic mulches break down with time and the original thickness typically reduces by 60 percent after a year. Organic mulches are mostly used for permanent crops or landscaped and noncrop areas, although they are also very effective for transplanted vegetables and garlic (Stone 2005b). Organic mulches are not effective against established perennial weeds such as field bindweed and nutsedge. Paper mulches work well for

transplanted vegetables, but proper laying and rapid crop establishment are critical if you want to avoid problems with the wind causing the mulch to rise up and damage the crop.

Organic mulches can be grown in place (figure 6.19), using cover crop varieties such as fava beans, vetches, or cereals such as oats, barley, and rye. The cover crops are planted on the top of the beds, grown to maturity, and killed by rolling or undercutting before the crop is planted (Stone 2005b). This technique is practical with strip tillage and plantings of large-seeded vegetables, but most often is used for transplanted vegetables that are planted with a mechanical transplanter designed to

Figure 6.19. No-till peppers planted into a cowpea mulch. The cowpeas were grown as a summer cover crop and cut at the soil surface to form a mulch into which the peppers were then directly transplanted. The cowpea mulch provides nitrogen and shades out weeds for most of the growing season. The mulch breaks down by the end of the season and does not interfere with pepper harvest. *Photo:* Milton E. McGiffen.

operate in high-residue situations. One risk of this technique is that if there is not sufficient cover crop residue to effectively inhibit weed emergence, weed control can be difficult, since effective cultivation is not possible.

BENEFICIAL ORGANISMS

Weeds suffer attacks from disease and insects just as crops do. For instance, additions of cover crop biomass to the soil can increase microbial (bacterial and fungal) activity, which may influence the survival of weed seedlings and degradation of the weed seed bank, although this effect is poorly understood. Biological control of weeds has been effective in rangeland situations but has been of little value for control of weeds in vegetables. Geese have been used for weed control in orchards, vineyards, and certain row crops. All types of geese will graze weeds. However, the small goose breeds such as Chinese weeder geese are considered best for row crops. Chinese weeder geese and other small breeds generally walk around delicate crop plants rather than over them. Geese prefer grass species and will eat other weeds and crops only after the grasses are gone. If confined, geese will even dig up and eat johnsongrass and bermudagrass rhizomes. Special care must be exercised when using geese: Avoid placing them near any grass crops (e.g., corn, sorghum, small grains, etc.), as grasses are their preferred foods. Tomatoes and other crops that produce fruit might also be vulnerable, especially when the fruit begin to color. At a certain point in the season, geese should be removed from tomato fields. Geese also require drinking water, shade during hot weather, and protection from dogs or other predators.

In organic farming operations, efforts are made to increase the level of organic matter in order to improve soil properties and fertility. There are some indications that additions of organic matter from cover crops and compost can reduce weed pressure. A 2-year study on California's Central Coast showed that organic amendments were associated with reduced shepherd's purse seed bank densities (Fennimore and Jackson 2003). It is unclear why organic matter may reduce weed emergence in this instance, but it is probably due to increased weed seed degradation by soil microbes.

CHEMICAL CONTROL

Herbicides are chemicals that kill or suppress plants by affecting plant physiological processes. There are few organically acceptable herbicides. Currently approved products contain acetic acid (vinegar), citric acid, or clove oil (eugenol), all of which disrupt membranes and cause leakage of fluids from cells. Ambient temperatures affect the efficacy of these materials, which work better in warmer temperatures. Organic herbicides are non-selective and can be used to kill small broadleaf weeds (i.e., ≤ 2 true leaves). However, once weeds get larger, organic herbicides are less effective on the weeds and must be applied in higher concentrations. Grass weeds are not controlled with currently available organic herbicides because the growing point is protected below the ground. Organic herbicides can be used to kill flushes of young weed seedlings of slow-germinating crops after the crops are planted but before they emerge. Onions appear to have some tolerance to postemergence over-the-top applications of low rates of some organic herbicides (Smith 2004b). The cuticle of the onion plant protects it from the herbicide's caustic action. Cutical development on onions may vary, though, depending on weather conditions. Before you spray a large field, it is wise to spray a small test area and examine it the next day to see if the onion leaves have sufficient cuticle to withstand the herbicide application.

HAND-WEEDING

Hand-weeding is generally necessary in organic vegetable production. The successful use of the other techniques described above can help make hand-weeding operations less tedious and more efficient.

NEW METHODS FOR THE FUTURE

Undoubtedly, more technological innovations will develop in the future to allow more mechanical weeding of crops. For instance, computer vision cultivators are being developed that can distinguish crop plants from weeds, and these will eventually be used to selectively remove weeds from the seed row. In addition, new developments in our understanding of soil microbiology may help us to better understand and possibly facilitate the biological control of weed seeds in the soil.

REFERENCES/RESOURCES

Bowman, G. 1997. Steel in the field: A farmers guide to weed management tools. Beltsville, MD: Sustainable Agriculture Network.

Boyd, N. S., and E. B. Brennan. 2006. Weed management in a legume-cereal cover crop with the rotary hoe. Weed Technology 20(3):733–737.

Brennan, E. B., and R. F. Smith. 2005. Winter cover crop growth and weed suppression on the Central Coast of California. Weed Technology 19(4):1017–1024.

Darlington, H. T., and G. P. Steinbauer. 1961. The eighty-year period for Dr. Beal's seed viability experiment. American Journal of Botany 48:321–325.

Davis, A. S., K. A. Renner, C. Sprague, L. Dyer, and D. Mutch. 2005. Integrated weed management: One year's seeding. East Lansing: Michigan State University Extension, Bulletin E-2931.

Elmore, C. L., J. J. Stapleton, C. E. Bell, and J. E. DeVay. 1997. Soil solarization: A nonpesticidal method for controlling disease, nematodes, and weeds. Oakland: University of California Division of Agriculture and Natural Resources, Publication 21377.

Fennimore, S. A., and L. E. Jackson. 2003. Organic amendment and tillage effects on vegetable field weed emergence and seedbanks. Weed Technology (17):42–50.

Grattan, S. R., L. J. Schwankl, W. T. Lanini. 1988. Weed control by subsurface irrigation. California Agriculture 42(3):22–24.

Ogbuchiekwe, E. J., and M. E. McGiffen, Jr. 2001. Efficacy and economic value of weed control for drip- and sprinkler-irrigated celery. HortScience 36:1278–1282.

Scopel, A. L., C. L. Ballare, and S. R. Radosevich. 1994. Photostimulation of seed germination during soil tillage. New Phytologist 126(1):145–152.

Shem-Tov, S., S. A. Fennimore, and W. T. Lanini, 2006. Weed management in lettuce (*Lactuca sativa*) with pre-plant irrigation. Weed Technology 20(4):1058–1065.

Smith, R. F. 2004a. Mustard cover crops for weed control in spinach and broccoli. University of California Cooperative Extension, Monterey County Crop Notes. July-August.

———. 2004b. Post emergence organic weed control in onions and broccoli. University of California Cooperative Extension, Monterey County Crop Notes. November-December.

Smith, R. F., T. Bensen. 2005. Precision cultivation evaluations. University of California Cooperative Extension, Monterey County Crop Notes. November-December.

Stone, A. 2005a. Weed 'em and reap, Part I: Tools for non-chemical weed management in vegetable cropping systems. DVD. Oregon State University, www.weedemandreap.org.

———. 2005b. Weed 'em and reap, Part II: Reduced tillage strategies for vegetable cropping systems. DVD. Oregon State University, www.weedemandreap.org.

Postharvest Handling for Organic Vegetable Crops

TREVOR SUSLOW

Optimal postharvest quality for organic vegetables—meaning that they possess the textural properties, sensory appeal, shelf life, and nutritional content desired by marketers, customers (resellers and restaurants), and consumers—is the combined result of careful implementation of recommended production inputs and practices, careful handling at harvest, and appropriate postharvest handling and storage. Some aspects of postharvest handling are common to both conventional and organic produce, while others are unique to organic vegetables or required if the vegetables are to be marketed as "registered organic" or "certified organic."

PLANNING FOR POSTHARVEST QUALITY

The effort that will culminate in an economic reward through the marketing of organic produce must begin well before harvest. Site selection and seed selection can be critical factors in determining the postharvest performance of any commodity. Individual cultivars vary in their inherent potential for firmness retention, uniformity of shape, characteristic color, disease and pest resistance, and shelf life—both in terms of postharvest visual quality and sensory quality (e.g., sugar-to-acid balance, aroma volatiles).

Organic vegetable producers, and direct marketers in particular, have traditionally included or focused on specialty and heirloom cultivars and on high-sensory-quality harvest strategies such as near full-ripeness on the vine. Interestingly, conventional markets have begun to recognize and respond to the expanding consumer demand for true heirloom and heirloomlike varieties, and these have begun to expand their share of the conventional produce market. Heirloom varieties and those selected for novelty, ethnic culinary, or flavor traits may be suitable for small-scale production and local marketing, while the same varieties might be a disastrous choice when the marketing plan includes shipment to more distant markets.

Some heirloom or specialty varieties are not well adapted for growth in the arid Mediterranean climate of California and tend to develop preharvest disorders or conditions that cause serious postharvest problems later on. Many vegetable varieties in this category simply do not hold up well under current postharvest handling and distribution regimes. The primary problems are bruising and cracking, compression damage in pallet loads, over-softening of the produce from water loss, and decay. In addition to genetic traits, environmental factors such as soil type, temperature, wind during fruit set, frost, and rainy weather at harvest can adversely affect storage life, suitability for shipping, and overall quality. In general, the varieties selected for conventional distribution and national organic distribution markets are those that are well suited to the growing region and season and can reach market without developing the potential defects described above.

Cultural practices in the field may have a dramatic impact on postharvest quality. For example, poor seedbed preparation for carrots may result in sunburned shoulders and green cores in many of the blunt-ended Nantes and other specialty carrots favored by consumers at farmers markets.

PLANNING FOR POSTHARVEST SAFETY

Planning for postharvest food safety should be part of any edible crop management plan. Organic vegetable production, regardless of the scale of operations, is not exempt from this fundamental regulatory responsibility, nor from related consumer expectations. Although food safety programs are not currently mandated by the U.S. Food and Drug Administration (FDA), the enforceable regulation is that growers and handlers must not distribute adulterated food to consumers. Practices that address many of these expectations are already embodied in an organic farm plan and organic handling system plan. Nonetheless, specific programs that are consistent with good agricultural practices (GAP) need to be developed and formalized for each crop and specific production field to minimize the risk of a variety of hazards or contaminants: chemical (e.g., heavy metals carryover), physical (e.g., sand and soil, wood, glass, plastic or metal shards), and biological (e.g., pathogenic *E. coli*, *Salmonella*, *Listeria*, parasites, and mycotoxins).

(Note: The Food Safety Modernization Act, signed into law in January, 2011, gives the FDA broader authority to regulate production and marketing of fresh produce. Details, beyond the scope of this chapter, are available online: www.fda.gov/Food/FoodSafety/fsma/ and www.ccof.org/foodsafety.php#overview.)

Prior land use, adjacent land use, water source and method of application, fertilizer choice (such as the use of manure), compost management, equipment maintenance, field sanitation, movement of workers between different operations, personal hygiene, domestic animal and wildlife activities, and other factors have the potential to adversely impact food safety. If you check the groups listed under "Other Resources" at the end of this chapter you can find information online for a number of these areas to help you develop a GAP program and on-farm self-audit. All growers, handlers, and marketers of fresh vegetables should be aware of the FDA's primary Guide to GAPs (www.cfsan.fda.gov/~dms), available in English, Spanish, Portuguese, French, and Arabic.

Many elements of a GAP plan are likely to be incorporated into a grower's basic organic crop management program and activities, but growers often find these measures also lead to improvements in efficiency and benefits to product quality following the implementation of a GAP and food safety system. Programs that a grower puts in place to ensure produce quality may also have a direct food safety benefit with minor modifications. Though this sort of thing is rarely documented, growers often see and share strong empirical evidence that the application of food safety programs have had a direct benefit to postharvest quality and market access.

Once prerequisite production and preharvest programs are in place, the grower must develop a systematic evaluation and implementation plan for GAP during harvest operations and any subsequent postharvest handling, minimal or fresh-cut processing, and distribution to customers and consumers. Considerations for these activities are covered briefly below.

HARVEST HANDLING

The inherent quality of fresh vegetables, like that of all produce, cannot be improved after harvest, only maintained for the expected window of time (shelf life) characteristic of the commodity. Part of what makes for successful postharvest handling is an accurate knowledge of how long this window of opportunity is, given your specific conditions of production, the season, your method of handling, and the distance to market. In some cases, the intended use for the produce and the customer's willingness to accept less-than-perfect quality may be additional factors that will lengthen this window. Under organic production, growers harvest and market their produce at or near peak ripeness more commonly than in many conventional systems. However, as discussed previously, organic producers often include more specialty varieties whose shelf life and resistance to shipping injury are reduced or even inherently poor. As a general approach, the following practices can help you maintain quality:

- Harvest during the coolest time of day to maintain low product respiration.

- Avoid careless and unnecessary wounding, bruising, compression, or crushing from overfilled or stacked harvest totes or packed cartons.

- Train harvest crews not to cause picking damage, such as fingernail injury or damage from pulling rather than twisting or snapping off fruit and vegetables.

- Shade the harvested product in the field to keep it cool. By placing harvest bins under a shade cloth structure or canopy or by covering stacked totes with an empty carton, clean fiberboard panel, reflective pad, or cloth, you greatly reduce heat gain from the sun, water loss, and premature senescence.

- If possible, move the harvested product into a naturally cool area, refrigerated storage facility, or postharvest cooling treatment as soon as possible. For some commodities—such as berries, specialty cucumbers and squash, tender greens, and leafy herbs—even 1 hour in the sun is too long.

- Do not compromise high-quality produce by mingling it with damaged, decayed, or decay-prone produce in a bulk or packed unit.

- Only use cleaned and, as necessary, sanitized harvest and trimming implements, equipment, and packing or transport containers.

These operating principles are important in all operations but carry special importance for many organic producers who have less access to postharvest cooling facilities.

POSTHARVEST COOLING FOR QUALITY

The control of temperature is the grower's single most important tool for maintaining the postharvest quality of produce. Some products are field cured (e.g., dry onions) or exceptionally durable (e.g., hard squash), but for those vegetables that do not fit one of those two categories the removal of field heat as rapidly as possible is highly desirable.

The act of harvesting cuts a vegetable off from its source of water, but the harvested vegetable is still alive and will continue to lose water, and therefore turgor, through the process of respiration. In terms of the postharvest biology of vegetables, respiration is an oxidative (oxygen-consuming) process that occurs within living plant cells and tissues to maintain the chemical energy supply and processes after harvest. The biological functions stimulated by wounding, natural maturation events, and continuing development after harvest are varied and complex. The harvested vegetable continues to draw from its stores of sugars, carbohydrates, proteins, lipids, and oils, resulting in the liberation of carbon dioxide, water, and heat. In low-oxygen conditions, respiration can become fermentative and produce carbon dioxide, ethanol, and other typically undesirable compounds that cause off-odors, off-flavors, or defects in the produce.

The main role of postharvest temperature management is to control the rate of respiration in order to slow ripening, control wound healing, modulate degreening (e.g., the formation of red pigment in tomatoes) to allow more extended storage before marketing, and delay the onset of physiological disorders and decay. Field heat can accelerate the rate of respiration, and with it the rate of quality loss. Proper cooling protects the produce's quality and extends its sensory (textural and taste) and nutritional shelf life.

The capacity to cool and store produce gives the grower greater market flexibility. Growers have a tendency to underestimate how much refrigeration capacity they will need to meet peak cooling demands. It is often critical that fresh produce rapidly reach its optimal pulp temperature for short-term storage or shipping if it is to maintain the most desirable combination of consumer market traits: visual quality, flavor, texture, and nutritional content. The considerations for selecting appropriate cooling methods and appropriate storage temperature and humidity conditions for a large diversity of vegetables are discussed in the two UC ANR publications mentioned below. The five most common cooling methods are

- *Room cooling*—an insulated room or mobile container with evaporative cooling (typically only used by very small operations) or equipped with a refrigeration unit or fully integrated cooling system. Room cooling is slower than other methods. Depending on the commodity, packing unit, and stacking arrangement, the product may cool too slowly to prevent water loss, premature ripening, or decay.

- *Forced-air cooling* (also referred to as pressure cooling)—fans used in conjunction with a cooling room to pull conditioned air through bins, totes, or packed units of produce. Although the cooling rate depends on the air temperature and the rate of airflow, this method is usually 75 to 90 percent faster than simple room cooling. Design considerations for a variety of small- and large-scale units are available in *Commercial Cooling of Fruits, Vegetables, and Flowers* (ANR Publication 21567).

- *Hydrocooling*—showering produce with chilled water to remove heat, and possibly to partially clean the produce at the same time. The use of an approved disinfectant in the water is essential, and some of the currently permitted products are discussed later in this chapter. Hydrocooling is not appropriate for all produce. It requires the use of waterproof containers or water-resistant waxed corrugated cartons. Currently, waxed corrugated cartons have limited options for recycling or secondary use, and reusable, collapsible plastic containers are gaining in popularity. Examples of vegetables suitable for hydrocooling are asparagus, sweet corn, peas, beans, cantaloupes, radishes, beets, and hardier greens. A more complete list of vegetables that are suitable for hydrocooling is available in *Postharvest Technology of Horticultural Crops: Third Edition* (ANR Publication 3311) as well as in *Commercial Cooling of Fruits, Vegetables, and Flowers*.

- *Top icing or liquid icing*—an effective method for cooling ice-tolerant commodities, and adaptable to both small- and large-scale operations. Ice-tolerant vegetables are listed in *Postharvest Technology of Horticultural Crops* and in *Commercial Cooling of Fruits, Vegetables, and Flowers*. To maintain the organic integrity of your produce, it is essential that you ensure that the ice is free of chemical, physical, and biological hazards.

- *Vacuum cooling*—a method that uses a vacuum chamber to cause the water within the plant to evaporate, removing heat from the tissues. This system works well for leafy crops that have a high surface-to-volume ratio, such as lettuce, spinach, and celery. The operator may spray water onto the produce before placing it into the vacuum chamber. As with hydrocooling, proper water disinfection is essential (see the section "Sanitation and Water Disinfection," below). The high cost of the vacuum chamber system restricts its use to larger operations. Many shippers of organic cool-season vegetables contract with a local cooling facility and make special arrangements for the required clean-out of chemical residues, segregated staging, and sequencing not commonly used for conventional loads. The spray-vacuum cooling of mixed conventional and organic loads is prohibited under the terms of organic certification.

In large cooling operations that handle both conventional and organic commodities, it is common to hydrocool (or water spray/vacuum cool) organic produce at the beginning of daily operation after a full cleaning of the facility and a complete water exchange. This practice is intended to prevent any carryover or cross-contamination onto organic produce from synthetic pesticides or other prohibited residues. Treatment so early in the morning generally requires at least overnight short-term storage of the produce. The injection of ozone into the cooling water stream has been shown to substantially oxidize trace pesticide residues that may remain in the water after it is used to cool nonorganic vegetables (a research example is provided in Wu et al. [2007]) .

Other postharvest issues that involve combined steps of unloading commodities from harvest bins, washing, and precooling must also be evaluated for adherence to organic standards. Some packinghouse operators use flotation as a way to move produce with a minimum risk of mechanical damage at the point of grading and packing. Entire bins are submerged in a tank of water treated with a chemical flotation aid that allows the picked produce to be gently removed and separated from the container. Lignin sulfonates are allowed in certified organic handling as flotation aids for water-based unloading of field bins or other density separation applications.

SANITATION AND WATER DISINFECTION

The sanitation of equipment and food contact surfaces and the disinfection of water should be integrated into every facet of postharvest handling for quality, optimal storage life, and food safety. Adequate washing and cleaning of produce (for those vegetables that can tolerate postharvest water contact), for the sake of customer presentation, decay and spoilage control, and reducing the risk of causing foodborne illness, should be considered by growers and handlers at all scales of production. Tolerant crops are often washed before shipping, especially if the produce is destined for extended storage or longer-distance distribution.

Besides reducing decay pathogens, a properly designed wash system in combination with approved antimicrobials can help prevent contamination by human pathogens including *Escherichia coli* (*E. coli*) O157:H7, *Salmonella, Shigella, Listeria,*

Cryptosporidium, Hepatitis virus, and *Cyclospora.* These and other pathogens have been associated with illnesses attributable to the consumption of domestic and imported fresh vegetables, though not necessarily from organic produce. Several cases of foodborne illness have been traced to poor or unsanitary postharvest practices, and especially to the use of nonpotable water for washing or cooling and the use of unsanitary shipping ice. In addition, though not extensively studied, pathogens such as *Campylobacter jejuni* have been isolated at very low frequency (approximately 1 to 3%) from diverse vegetables taken from farmers market stands, though the data do not specifically link them to certified organic producers.

For organic handlers, the nature and prior use of cooling water is a special consideration. Postharvest water cannot at any time contain prohibited substances in dissolved form. Responsibility for this is an important issue all along the way for the organic producer as well as the handler, the processor, and the retailer. Even incidental, unintentional contamination with a prohibited material would keep the product from being certified organic. Organic producers, packers, and handlers are required to keep accurate, specific records of postharvest wash or rinse treatments, with all materials identified by brand name and source. For a more complete discussion of water disinfection, see *Postharvest Chlorination* (ANR Publication 8003) and *Making Sense of Rules Governing Chlorine Contact in Postharvest Handling of Organic Produce* (ANR Publication 8198).

Put briefly, the proper use of a disinfectant in postharvest wash and cooling water can help prevent postharvest diseases and foodborne illnesses. Because most municipal water supplies are chlorinated and the importance of water disinfection is well recognized, organic growers, shippers, and processors may use chlorine within specified limits. All forms of chlorine (including liquid sodium hypochlorite, granular calcium hypochlorite, and chlorine dioxide) are restricted materials as defined by existing organic standards. Any application of chlorine must conform to the maximum residual disinfectant limit under the Safe Drinking Water Act, currently 4 mg/L (4 ppm) expressed as Cl2. The California Certified Organic Farmers (CCOF) regulations have permitted this threshold of 4 ppm

residual free chlorine, measured downstream of the product wash. Other certification agents may interpret or apply this restriction in different ways, so make sure to check with your certifying agent before you use chlorine.

For both organic and conventional operations, liquid sodium hypochlorite is the most commonly used form of chlorine. For optimal antimicrobial activity with a minimal concentration of applied hypochlorite, make sure the pH of the water is adjusted to between 6.5 and 7.0. At this pH range, most of the chlorine takes the form of hypochlorous acid (HOCl), which delivers the highest rate of microbial kill and a negligible release of irritating and potentially hazardous chlorine gas (Cl_2). Chlorine gas will only exceed safe levels if the water is too acidic—below pH 3.5. This generally only results from a sudden influx of acid and is easily prevented with proper injection or other dosing management. The common irritating off-gassing associated with hypochlorite most typically comes from chloramines, a form of combined chlorine. Monochloramines are abundant when pH goes above 8.3. If you maintain a pH of around 7, you will keep about 80 percent of the chlorine in the hypochlorous acid (active) form with very little chloramine release, and have no need to be concerned over hazardous gas formation.

For organic growers, products used for pH adjustment must also come from a natural source (for example, citric acid, sodium bicarbonate, or vinegar). Calcium hypochlorite, properly dissolved, may also reduce sodium injury to sensitive crops (e.g., light-skinned cucumbers and immature yellow squash), and limited evidence points toward extended shelf life for tomatoes and bell peppers resulting from calcium uptake. Useful guides, tables, and links to sites that will help you determine the amount of sodium or calcium hypochlorite to add to clear, clean water for disinfection are available from several sources including

• UC Postharvest Technology, Research, and Information Center (http://postharvest.ucdavis.edu)

• UC Vegetable Research and Information Center (http://vric.ucdavis.edu/veginfo/veginfor.htm)

• UC Small Farm Center (www.sfc.ucdavis.edu)

• ATTRA—National Sustainable Agriculture Information Service (http://attra.ncat.org/)

As a general practice, you should keep field soil to a minimum on product, bins, totes, and pallets by dry brushing the harvest containers and pallets and prewashing the produce. This will significantly reduce the need for disinfectant in the water and will lower the total required volume of antimicrobial agents. Prewashing also removes plant exudates released from harvest cuts or wounds, which can otherwise react rapidly with oxidizers such as hypochlorite and so require higher rates of the chemical to maintain the target of 4 ppm downstream activity. Oxidizers such as ozone are highly reactive with plant exudates and interfere with effective dose maintenance.

RECOMMENDED STEPS TO OPTIMIZE POSTHARVEST CHLORINATION

- Minimize all sources of chlorine demand (soil, plant debris, heavily wounded or decayed produce) on incoming product.

- Inspect incoming product during precooling staging for excessive amounts of adhering soil and non-product plant material (such as leaf or vine trash) on totes, cartons, and pallets. Remove as practical.

- Provide feedback as needed to harvest operations or crews to help them improve performance in reducing sources of chlorine demand at the field level.

- For vegetable products that will tolerate it, light mechanical cleaning (such as dry brushing or brush washing) may significantly reduce chlorine demand and extend the clarity of the process water and the functional disinfection activity.

- If using chlorine or hypochlorite, manage and monitor postharvest water to maintain a pH of 6.5 to 7.0 and a level of free chlorine necessary to achieve disinfection goals while not exceeding 4 ppm, measured at the point the product is removed from that operational step. For example, adjust the chlorine level in dump and flume tank injection water to maintain 50 ppm, and maintain the chlorine level in recirculating water at 4 ppm or less at the return sump (the farthest point from injection, where product transfers to a lift conveyor).

- For products that will tolerate it, such as tomato and melons, heating chlorinated water increases

the effectiveness of disinfection. The benefits of this practice must be balanced, however, with the reduced stability and increases in irritating off-gassing that also come with heating. Heating the initial receiving water above the temperature of the incoming product (field heat) also minimizes the potential for infiltration of water into the product.

- If necessary, add approved flocculants to capture suspended sediments in the water and hold the water in a retention sump basin. Screen and filter recirculating water to reduce chlorine demand and improve performance of whatever oxidizer (e.g., ozone and peroxides) or nonoxidizing disinfectant (e.g., UV systems and plant essential oils) you use.

- The dump tank, flume, and hydrocooler sump basins and any sediment collection points in equipment must be cleaned daily. Sediments commonly serve as reservoirs for decay pathogens and pathogens of concern for human food safety.

- Develop a system of periodic partial or full turnover to clean the water, balancing the cost and time delays of cooling or heating postharvest water against the goal of minimizing the turbidity and electrical conductivity (salt buildup) of the water. Develop a simple rating system for assessing turbidity thresholds for periodic water exchange. A simple turbidity tube with standard Secchi disc (black and white pattern) is very affordable and should be available from a water quality supply company.

- Ensure that responsible personnel are trained in the function and operation of equipment and monitoring kits. Define the timing, frequency, and procedures to be used in monitoring in a standard operating procedure for each product line.

- Ensure that responsible personnel are trained to implement and document any corrective actions that must be taken to meet product quality and safety needs within NOP standards.

Nonchlorine Oxidizers for Water Disinfection

Ozone is an attractive alternative to chlorine for water disinfection and other postharvest applications. Ozonation is a powerful oxidizing treatment, effective against chlorine-tolerant decay microbes (such as some *Fusarium* and *Alternaria* spore forms) and foodborne pathogens, especially parasite cysts and viruses, and it acts far more quickly than permissible concentrations of chlorine. Speed may be a distinct advantage for a cooling or wash procedure with a short contact time. Ozone oxidative reactions create far fewer disinfection by-products (e.g., trihalomethanes, which are a health and environmental concern) than chlorination. You may decide to use ozonation rather than chlorination in your organic postharvest operation either as a matter of preference, as a way to meet the requirements of certain markets that do not allow chlorine contact with foods, or as a way to appeal to particular consumer sectors. Be aware that the capital and operating costs for ozone treatment typically are higher than for chlorination or other available methods.

Ozone must be generated on-site at the time of use and is very unstable, lasting as little as 20 minutes even in clear water, and leaves no residual material. Its rapid breakdown to oxygen makes ozonation a desirable treatment, but it also means that maintaining effective levels of ozone in large volumes is very challenging and rarely practical. Ozone systems are being used by vegetable packers and fresh-cut processors, conventional and organic, and typically at a focused point in the process. Ozonation for the final rinse water has become fairly common. Clear water is essential for optimal performance, and adequate to superior filtration of the input or recirculating water is an absolute requirement. Depending on the scale of the operation and the volume of ozone you will need to generate, complete-system start-up costs begin at about $10,000. Small-scale units available for a few thousand dollars are suitable for situations with limited water use and small batches of produce. For specifications and installation information, consult an experienced ozone service provider. Networking with growers and packers who have installed ozone systems is an effective way to find a good ozone generator supplier from among the many. You want an outfit that will provide a high level of system design and technical operation service as a package, rather than one that will simply ship a unit for installation.

Food-grade hydrogen peroxide (0.5 to 1%) and peroxyacetic acid are other options. In general, peroxyacetic acid (PAA, 11% hydrogen peroxide and 15% acetic acid) has good efficacy in water dump tanks and water flume sanitation applications. PAA performs very well compared to chlorine and ozone at removal and control of microbial biofilms (tightly adhering slime) in dump tanks and flumes. PAA formulations have a higher per-unit cost than hypochlorite. Materials that are approved for organic uses are available, and marketers have recently started providing both large, bulk-volume units and smaller-volume units for ease of handling in smaller operations.

A number of other postharvest treatments are allowed for organic produce. Examples include natural sources of organic acids (e.g., acetic acid vinegar or malic acid), spice extracts and plant essential oils (e.g., rosemary, thymol, clove, spearmint, peppermint, or cinnamaldehyde), thiosulfinates (allicin or garlic extract), and copper ions.

CLEANERS, SANITIZERS, AND DISINFECTANTS

A partial list of allowed cleaners, disinfectants, sanitizers, and postharvest aids follows.

- *Acetic acid:* allowed as a cleanser or sanitizer. The vinegar used as an ingredient must be from an organic source.

- *Alcohol (ethyl):* allowed as a disinfectant. Alcohol must be from an organic source.

- *Alcohol (isopropyl):* may be used as a disinfectant under restricted conditions.

- *Ammonium sanitizers:* quaternary ammonium salts are a general example in this category. Quaternary ammonium may be used on non-food-contact surfaces. Its use is prohibited on food-contact surfaces except on specific types of equipment where alternative sanitizers significantly increase the corrosion of the equipment. Detergent cleaning and rinsing procedures must follow quaternary ammonium application. Monitoring must show no detectable residues prior to the start of organic packaging (e.g., for fresh-cut salads).

- *Bleach:* calcium hypochlorite, sodium hypochlorite, and chlorine dioxide are allowed as sanitizers for water and food-contact surfaces. In California, product (fresh produce) wash water treated with chlorine compounds as a disinfectant cannot exceed 4 ppm residual chlorine measured downstream of product contact. Higher levels of bleach are permitted in the makeup water.

- *Detergents:* allowed as equipment cleaners. This category also includes surfactants and wetting agents. Products must be evaluated on a case-by-case basis.

- *Hydrogen peroxide:* allowed as a water and surface disinfectant.

- *Ozone:* considered *GRAS* (generally regarded as safe) for produce and equipment disinfection. Exposure limits for worker safety do apply.

- *Peroxyacetic acid:* water disinfectant and fruit and vegetable surface disinfectant.

OTHER POSTHARVEST TREATMENTS

There are three additional postharvest treatments that you can use on organic produce:

- *Carbon dioxide:* permitted for postharvest use in modified- and controlled-atmosphere storage and packaging. This includes modification of nitrogen and oxygen levels by means of differential membrane filtration and introduction from compressed gas cylinders. For those crops that can tolerate treatment with elevated CO_2 levels ($\geq 15\%$), this treatment can suppress decay and control insect pests.

- *Fumigants:* allowed if the fumigant chemicals are naturally occurring forms (e.g., heat-vaporized acetic acid). Materials must be from a natural source.

- *Wax:* must not contain any prohibited synthetic substances. Acceptable sources include carnauba or wood-extracted wax. Products that are coated with approved wax must be labeled as such on the shipping container.

IMPORTANCE OF OPTIMAL STORAGE AND SHIPPING TEMPERATURES

Although we stress the rapid and adequate cooling as a fundamental element in postharvest handling, many vegetables are subtropical in origin and so are susceptible to chilling injury. Chilling injury occurs when sensitive crops are exposed to temperatures that are low but still above the freezing point. Damage can often be induced by a very brief exposure to cold temperatures, and it may not become apparent until several days later or until the harvested produce is transferred to warmer conditions at the retail market. Some examples of

chilling-sensitive crops are basil, tomato, eggplant, green beans, okra, and yellow crookneck squash. Different parts of some vegetables have distinct sensitivities. In eggplant, the cap or calyx is more sensitive to chilling and turns black before the fruit itself is affected. The effects of chilling injury are cumulative in some crops.

Damage from chilling injury may not be apparent until the produce is removed from low-temperature storage. Depending on the duration and severity of the chilling, symptoms may become evident in the following ways several hours or a few days after the produce is returned to warmer temperatures:

- pitting and localized water loss

- browning or other skin blemishes

- internal discoloration

- increased susceptibility to decay

- failure to ripen or uneven color development

- loss of flavor, especially involving characteristic volatiles

- development of off-flavors

Temperature management also plays a key role in limiting water loss in storage and transit. As the primary means of lowering respiration rates of vegetables and fruit, reduced temperature has an important relationship to relative humidity and thus directly affects the product's rate of water loss. The cooler the air is, the less moisture it is capable of holding. In a storage facility, the relative humidity of the ambient air in relation to the relative humidity of the produce (essentially 100%) directly influences the rate of water loss from the produce at any point in the marketing chain. Water loss may result in wilting, shriveling, softening, browning, stem separation, or other defects.

Transporting produce to and displaying it in roadside stands or farmers markets can often mean extended periods of exposure to direct sunlight, warm (or even high) temperatures, and low relative humidity. Rapid water loss under these conditions can result in limp, flaccid greens and a loss of appealing natural sheen or gloss in fruits and vegetables. By providing postharvest cooling before and during transport and a shade structure during display, you can minimize rapid water loss at market outlets.

WAXES AND PACKAGING

Approved vegetable waxes or sheens are effective tools for reducing water loss and enhancing produce appearance. Waxes for organic use may not contain synthetic substances. Carnauba and other natural waxes are acceptable. Waxed products must be labeled. Uniform application and coverage are important when you are using waxes or natural plant oils, so you need to use proper packing line brushes or rolling sponges.

Plastic wraps or other food-grade polymer films also retard water loss. Adequate oxygen exchange is necessary to prevent fermentative respiration and the development of ethanol and off-odors or off-flavors. Wraps or bags must have small perforations or slits to prevent these conditions, especially when temperature management is unavailable. The exposure of bagged or tightly wrapped produce to direct sunlight will cause the product's internal temperature to rise rapidly. Water loss will result and, with exposure to cooler conditions in storage, transport, or outdoor storage under night-time temperatures, free water will condense from the saturated vapor and lead to accelerated decay.

Proper packing design and packaging can also minimize water loss. To minimize condensation inside a bag and reduce the risk of microbial growth, bags may be vented, microperforated, or made of a material that is permeable to water vapor. Barriers to water loss may also function as barriers to cooling, however, so packing systems should be carefully selected for the specific application with this in mind. Packaging materials, storage or transport containers, or bins that contain synthetic fungicides, preservatives, or fumigants (or any bags or containers that have previously been in contact with any prohibited substance) are not allowed for organic postharvest handling. In small-scale handling, the reuse of corrugated fiberboard containers from conventional produce is strongly discouraged by organic certifying organizations. Reuse of hard-to-clean containers that have held conventional produce may even be prohibited by specific organic registration or certifying authorities.

During transportation and storage, relative humidity (or, more properly, vapor pressure deficit) is critical, even at low temperatures. For a more complete discussion of optimal relative humidity for fruits and vegetables and the principles for prevention of water loss, see *Commercial Cooling of Fruits, Vegetables, and Flowers* (ANR Publication 21567).

Specialized films that create modified atmospheres (MA) when sealed as a bag or pouch are available for many produce items that have well-characterized tolerances for low oxygen and elevated carbon dioxide levels. Not all commodities benefit from MA.

ETHYLENE

The management of ethylene may be another postharvest consideration for quality maintenance during storage and transportation. Ethylene, a natural hormone produced by plants, is involved in many natural functions during development, including ripening. Ethylene treatments may be applied to degreen fruits or accelerate ripening events in fruits harvested at a mature but unripe stage. For a detailed discussion of the role of ethylene in ripening and postharvest management, see *Postharvest Technology of Horticultural Crops* (ANR Publication 3311).

In organic handling, ethylene gas produced by catalytic generators is prohibited for all vegetable products. Ethylene generators have recently been allowed for tropical fruits but not for subtropicals such as citrus and tomatoes. In response to customer preference and expectations, the majority of ethylene-responsive organic produce is harvested nearly or fully ripe, so this restriction on ethylene use does not currently constitute a significant problem.

In contrast to its role in ripening, ethylene from plant sources or environmental sources (e.g., the combustion of propane in lift trucks) can be very damaging to sensitive commodities. Ethylene-producing equipment and other ethylene sources should not be stored with fruits or vegetables that are sensitive to it. External ethylene will stimulate a loss of quality, reduce shelf life, increase disease, and induce specific symptoms of ethylene injury, such as the following:

- russet spotting of lettuce

- yellowing or loss of green color (as in cucumber, broccoli, kale, and spinach)

- increased toughness in turnips and asparagus spears

- bitterness in carrots and parsnips

- yellowing and abscission (dropping) of leaves in brassicas

- softening, pitting, and development of off-flavor in peppers, summer squash, and watermelons

- browning and discoloration in eggplant pulp and seeds

- discoloration and off-flavor in sweet potatoes

- increased ripening and softening of mature green tomatoes

Besides providing adequate venting or fresh air exchange to keep ethylene levels under control, you can use ethylene adsorption or conversion systems that are designed to prevent damaging levels (as low as 0.1 ppm for some items) from accumulating in storage and during transportation. Potassium permanganate ($KMnO_4$) air filtration systems or adsorbers are allowed for organic postharvest handling, provided that strict separation is ensured to prevent product contact. For greatest efficacy, room air must be circulated through these filters.

Other air filtration systems available for ethylene removal in cold rooms are based on the circulation of room air through units containing glass rods treated with a titanium dioxide catalyst and ultraviolet light inactivation. Corona discharge or ultraviolet light–ozone-based purification systems for both ethylene elimination and disinfection of airborne spores are commercially available.

SPECIAL ISSUES

Although irradiation technologies are strongly disfavored by much of the organic food industry, X-ray irradiation is allowed for metal detection in packaged products. Metal detection is a common practice in many minimally processed and packaged organic vegetables and salad mixes.

The use of incompletely composted animal manure in organic production is prohibited due to postharvest food safety concerns. Organic standards specify a waiting period of 60 to 120 days, depending on certifying agency and crop, between the date the composted animal manure is applied to the soil and the date a crop intended for human consumption is planted. The California Certified Organic Farmers organization (CCOF) requires that all animal manure used for soil amendment be composted or treated according to current standards for Class A level pathogen reduction, as specified by the U.S. Environmental Protection Agency (EPA). Documentation of the compost process conditions for each batch is an essential part of the required record keeping that ensures compliance with preventive food safety programs. In addition, composting reduces the potential for inhibition of plant growth that is often associated with the use of raw manure.

Properly composted manure can be applied directly to growing vegetable crops with little concern. Ensuring adequate and thorough composting is not something to be taken for granted and requires stringent process control and deliberate measures to prevent recontamination of stabilized compost. However, although composting can degrade many if not most organic contaminants (i.e., pesticides), it cannot eliminate heavy metals. The composting process concentrates heavy metals that are a particular concern with sewage sludge (biosolids), a composted product occasionally used in production that can impact postharvest safety. The use of biosolids is prohibited by many organic certification organizations, including the CCOF.

Shippers must be aware of special requirements for transporting organic produce, whether by highway truck, air carrier, or containerized marine and intermodal shipping. Mixed-load shipment of organic and conventional products are permitted if "Organic" labeling is prominently and clearly displayed. In addition, there must be no risk that organic commodities will be contaminated by or come into direct contact with conventional product. Typically, carriers of bulk, raw organic product must maintain complete records of clean-out dates, detergents, and sanitizers used. Procedures for transport carrier cleaning and other treatments must include steps to prevent contamination from cleaners or fumigants, ripening agents, pest control agents, diesel fumes, and vehicle maintenance products.

RETAIL HANDLING, STORAGE, AND DISPLAY

Retailers are not required to be certified for organic handling but have adopted guidelines to do the following:

- prevent commingling with non-organic-certified products

- prevent contamination with prohibited substances

- keep records of integrity through delivery to the customer

Here is a list of key elements of retail handling and display of organic produce:

- Organic produce is palletized and stored separately.

- Produce that requires washing is washed in designated, sanitized basins and placed in designated totes, bins, or racks.

- Trimming is performed with tools designated for use only on organic foods.

- Organic items are clearly marked and displayed to avoid commingling, customer confusion, or contamination.

- Bulk organic greens are stored in separate bins with separate tongs or scoops.

- Produce staff are trained regarding procedures for organic products.

- Specific and appropriate pest control procedures are employed.

REFERENCES/RESOURCES

Gross, K. C., C. Y. Wang, and M. Saltveit. 2004. The commercial storage of fruits, vegetables, and florist and nursery stocks. USDA Agricultural Handbook #66. http://www.ba.ars.usda.gov/hb66/contents. html.

Kader, A. A., tech. ed. 1992. Postharvest technology of horticultural crops, 3rd ed. Oakland: University of California Division of Agriculture and Natural Resources, Publication 3311.

Ryall, A. L., and W. J. Lipton. 1983. Handling, transportation, and storage of fruits and vegetables. Westport, CT: AVI Publishing.

Suslow, T. 1997. Postharvest chlorination: Basic properties and key points for effective distribution. Oakland: University of California Division of Agriculture and Natural Resources, Publication 8003.

Suslow, T. 2006. Making sense of rules governing chlorine contact in postharvest handling of organic produce. Oakland: University of California Division of Agriculture and Natural Resources, Publication 8198.

Thompson, J. F., P. E. Brecht, T. Hinsch, and A. A. Kader. 2000. Marine container transport of chilled perishable produce. Oakland: University of California Division of Agriculture and Natural Resources, Publication 21595.

Thompson, J. F., F. G. Mitchell, T. R. Rumsey, R. F. Kasmire, and C. H. Crisosto. 1998. Commercial cooling of fruits, vegetables, and flowers. Oakland: University of California Division of Agriculture and Natural Resources, Publication 21567.

Wu, J., et al. 2007. Removal of residual pesticides on vegetables using ozonated water. Food Control 18(5):466–475. http://www.sciencedirect.com/science/journal/09567135.

INTERNET RESOURCES

ATTRA (National Sustainable Agriculture Information Service), http://attra.ncat.org.

California Certified Organic Farmers, Certification information, www.ccof.org/certification.php.

Community Alliance with Family Farmers (CAFF), www.caff.org.

Organic Materials Review Institute, www.omri.org.

National GAPs Program Web Links, www.gaps.cornell. edu/gapsd/Weblinks.html.

U.S. Food and Drug Administration, Center for Food Safety and Applied Nutrition, www.fda.gov/Food/ResourcesForYou/Consumers/ucm114299.htm.

UC GAPs Program, http://ucgaps.ucdavis.edu.

UC Postharvest Technology Research and Information Center, http://postharvest.ucdavis.edu.

UC Small Farm Program, http://sfp.ucdavis.edu.

PHOTO: KATHY KEATLEY GARVEY

■ Index

Note: Page numbers in **bold type** indicate major discussions. *Italic* type is used to indicate tables, e.g., 15*t*, and figures, e.g., 54*f*.

G

GAP (good agricultural practices), and food safety, 68

garlic
  nitrogen requirements of, 32t
  weed control in, 52, 59, 62

garlic rust fungus, 40

geese, for weed control, 63

goals, in business plan, 13

good agricultural practices (GAP), and food safety, 68

grass weeds, control of, 60, 63, 64

gray mold, control of, 43

greenhouse crops, sanitation measures, 42, 45

greens
  nitrogen requirements of, 32
  postharvest cooling of, 70

groundsel, control of, 49

growth, organic sales (California), 19

Guide to GAPs, from FDA, 68

gypsum, 30, 35

H

hand-weeding, 49, 50f, 52, 64

harvesting practices, 67, 68–69

heirloom varieties, postharvest problems, 67

Hepatitis virus, 71

herbicides, 64

horticultural oils, 43

humus, 29

hydrocooling, postharvest, 70

hydrogen peroxide
  for disease control, 43
  for water disinfection, 73, 74

I

ice, quality of, 70, 71

IFOAM (International Federation of Organic Agricultural Movements), 4, 5

information, sources of, 15t. *See also resources at end of each chapter*

inputs. *See* materials, compliant and noncompliant

inspections, certification process, 4, 7–8

interest on operating capital, 20

International Federation of Organic Agricultural Movements (IFOAM), 4, 5

irradiation, 76

irrigation
  disease control factor, 44, 45
  nitrogen loss and, 32
  weed control and, 49, 50–51

J

johnsongrass, 63

K

kale, ethylene injury of, 75

kaolin, 43

kelp, 33t, 35

knives, cultivation, 55, 56f, 57

L

labeling regulations, history, 1

labor costs, estimating, 20, 21t

lambsquarters, 59

land use verification, 3, 6, 7

leaching, 31–32, 33

leafhoppers, vectors of disease, 41

legumes
  as cover crop, and weed pressure, 53
  nitrogen fixation, 32

lettuce

disease control in, 40–41

economics of, sample enterprise, 19–26

ethylene injury of, 75

nitrogen requirements of, 32t

postharvest cooling of, 70

weed control for, 49, 50, 51f, 57, 58t

lettuce corky root disease, 44

lettuce drop, 40t, 44

Lettuce mosaic virus, 40, 44

Lettuce necrotic stunt virus, 42

lignin sulfonates, 70

lime
  for disease control, 42
  fertilizer, 35

liquid icing, postharvest cooling, 70

liquid tea fertilizers, 34–35

Listeria, 68, 70

little mallow, control of, 49

M

MA (modified atmosphere) films, 75

magnesium (Mg), 29, 30

Making Sense of Rules Governing Chlorine Contact . . . , ANR publication, 71

management, section of business plan, 12t, 13, 15–16

manure
  as fertilizer, 33t, 34
  food safety and, 76
  weed seeds in, 52

marketplace analysis, 12t, 13, 14

market segments, 14

materials, compliant and noncompliant, 8–9. *See also certification, as organic*